Marketing Planning for Culture and the Arts

Library and Archives Canada Cataloguing in Publication
Colbert, François
Marketing planning for culture and the arts / François Colbert ... [et al.].
Includes bibliographical references and index.
ISBN 978-2-9808602-2-5

1. Arts–Marketing. 2. Cultural industries–Management. 3. Marketing.
I. Colbert, François II. Title.

NX634.M38 2008 700.68'8 C2008-903760-X

Printed in Canada

Managing editor: Louise St-Pierre
Copy editing: Jane Broderick
Cover design: Stéphane Lortie, Atelier de graphisme, HEC Montréal
Layout: Catherine Lafortune-Rabbat, Atelier de graphisme, HEC Montréal

Distributed by the Carmelle and Rémi Marcoux Chair in Arts Management, HEC Montréal,
3000, chemin de la Côte-Sainte-Catherine, Montréal (Québec) H3T 2A7 Canada
tel.: 514 340-5629, www.gestiondesarts.com

Legal Deposit – Bibliothèque nationale du Québec – National Library of Canada, 2008

Bibliography

Johanne Brunet, MBA, PhD (Industrial and Business Studies), University of Warwick (UK), is Associate Professor in Marketing at HEC Montréal. She specializes in international marketing, strategic marketing, creativity and innovation. She is an Associate Member of the Carmelle and Rémi Marcoux Chair in Arts Management. She has been involved in cultural industries for over 25 years. She has served on the boards of numerous organizations, including the Canadian public broadcaster, CBC/Radio-Canada.

François Colbert, CM, MBA, MSc, is Professor of Marketing at HEC Montréal, where he holds the Carmelle and Rémi Marcoux Chair in Arts Management and is responsible for the graduate diploma program in management of cultural organizations. He is Editor of the *International Journal of Arts Management.* Professor Colbert has been actively involved in the field of arts and culture for over 30 years. He is the author of more than 150 publications on issues related to arts management. In 2002 he was awarded the Order of Canada for his many achievements and for his unique contributions in developing the field of arts management.

Dan J. Martin is Director of the Institute for the Management of Creative Enterprises at Carnegie Mellon University. He served in senior management positions at CSC Repertory (New York City), the Virginia State Company (Norfolk) and the Walnut Street Theatre (Philadelphia) before joining the academic community. He also consults with cultural organizations in strategic planning, information technology and finance management. He has taught classes and led workshops on cultural management issues in the United States, Canada, Germany, Spain, Austria, Greece and Italy.

Jennifer Radbourne is Head of the School of Communication and Creative Arts at Deakin University in Melbourne, Australia. Her teaching and research have focused on arts management and arts marketing and she has published extensively in these areas internationally. Professor Radbourne serves on the editorial boards of three international journals and on the Scientific Committee of AIMAC (International Association of Arts and Cultural Management). She has consulted with local and state governments and arts organizations in Australia and internationally, in strategic and marketing planning and business development for the arts and creative industries.

Philippe Ravanas, MBA, is Marketing Professor in the Graduate Program of the Arts, Entertainment and Media Management Department at Columbia College Chicago. He is past Vice President of Corporate Communications for Euro Disney in Paris and past Manager of Client Development for Christie's in London. He is a Registered Expert and Advisor for the United Nations Conference on Trade and Development and for the World Trade Organization.

J. Dennis Rich is Chairperson of the Arts, Entertainment and Media Management Department at Columbia College Chicago. He is a founder of the Center for Arts Policy at Columbia College. He is also a consultant, researcher and lecturer. His international activities include colloquia and workshops for the United States Information Agency in India and lecture series in Mexico, Russia, Austria, Hungary, Germany and France. He trained in theatre, has an extensive arts management background and has managed a variety of performing arts organizations.

Introduction

By François Colbert and Dan J. Martin, Editors

No established recipes exist for effective marketing. There are no magic formulas for success in generating large, enthusiastic and engaged audiences. If there were an algorithm or guaranteed process for attracting the attention of prospective patrons, for generating their interest, for motivating them to make a decision and, ultimately, to take action, there would be far fewer books on marketing on the bookshelves. Cultural organizations would be presenting their work to capacity audiences! Successful marketing effectively and efficiently connects target markets with the work being presented or exhibited; therefore, the process is highly specialized, with specific messages directed to *market segments* based on where those segments – and sometimes even individual consumers within those segments – are situated within the *purchasing process.* Marketing techniques cannot be copied wholesale from one program or service to another (not even within the same organization) or from one institution to another. Each cultural organization and each of the programs it offers presents unique marketing challenges and opportunities; the same is true for the market segments the institution hopes to engage for its programs. The marketing successes of peer institutions may offer *examples* of how to solve a marketing challenge, but the recipe they have developed and refined will be ineffective in a different setting, at a different time and with a different set of stakeholders.

If there is any doubt about the infinite variety of marketing tools and techniques available, one need only browse bookstore shelves, trade magazines, scholarly journals and Web sites for advice on the latest way to attract those hard-to-reach potential arts patrons. New suggestions for engaging today's elusive youth market, the middle-aged urban pioneer, or the currently disenfranchised or discouraged segment emerge regularly. What appears to frustrate many committed cultural marketing professionals, however, is the lack of a clear process for taking established marketing theory and principles – the fundamentals that have *not* changed in spite of the evolution in communication systems – and building a strong, *practical* marketing plan, an operating plan based on those principles, and a unified and cohesive set of strategies and objectives.

The Structure

Marketing Planning for Culture and the Arts is intended to provide the fundamentals. It is not a book about the latest marketing strategies and tactics – though we do discuss the integration of information technology into the communication process – but, rather, an explanation of how to structure and coordinate those strategies and tactics. The reader is expected to have a basic knowledge of marketing theory, but theoretical proficiency is not required to make good use of the material contained in the book.

Marketing Planning for Culture and the Arts is presented in two parts:

Part One

1. Elements of Arts and Culture Marketing
2. The Marketing Planning Process
3. Integrating Information Technology into the Marketing Process

Part Two

4. Planning for Fundraising and Sponsorship
5. Planning for International Touring in the Performing Arts
6. International Marketing for Cultural Industries (Film, Publishing, Recording)

Why Part Two? you may ask. Planning for Fundraising and Sponsorship is included because of the role it plays in the communication process between cultural organizations and their audiences. Fundraising messages can enhance or negate the efforts of the Marketing Department; they must be integrated and complementary. The challenges of both Planning for International Touring in the Performing Arts and the basics of International Marketing for Cultural Industries include eccentricities within the marketing process that warrant special notice and discussion.

The Scope

Speaking of eccentricities, the theories, principles and practices discussed in *Marketing Planning for Culture and the Arts* are *not* exclusive to performing arts versus visual arts organizations or to not-for-profit versus for-profit creative enterprises. There is no difference in the fundamentals that drive marketing planning except, obviously, the consumer perspective: For-profits are market- or consumer-driven while not-for-profits are product-driven but market-sensitive. The only other differences are the strategy variables, including time, capacity and distribution. Performing arts organizations benefit from the motivating forces of a specific curtain time or performance date for the audience to experience the work, or the exclusivity of participation based on limited capacity. But these are superficial differences. While most visual arts organizations may not be able to exploit these variables in the same way as performing arts organizations, they can derive marketing benefits from them by highlighting "flexibility" of attendance – patrons' freedom to attend when it is convenient for them to do so. Our point? It should be clear that these principles and practices transcend specific sectors; they can be translated to and implemented by any creative enterprise.

The Contributors

The six people who have contributed to this book share strongly held beliefs about marketing that will become apparent as you read. There are, however, some differences of opinion. This is by design. These varied voices are intended to provide you with an opportunity to choose among the different options and to develop a menu of strategies and practices that will best serve your organization and its mission. You also will encounter some repetition across the chapters. This is also by design. It is intended to reinforce important concepts and to illustrate the implementation of marketing principles in a variety of situations.

Contents

ELEMENTS OF ARTS AND CULTURE MARKETING

THE MARKETING PLANNING PROCESS

INTEGRATING INFORMATION TECHNOLOGY INTO THE MARKETING PROCESS

PLANNING FOR FUNDRAISING AND SPONSORSHIP

PLANNING FOR INTERNATIONAL TOURING IN THE PERFORMING ARTS

INTERNATIONAL MARKETING FOR CULTURAL INDUSTRIES (FILM, PUBLISHING, RECORDING)

PLAN

Chapter 1

Elements of Arts and Culture Marketing

By François Colbert

OBJECTIVES

- Introduce the main marketing concepts that are applied to arts marketing
- Understand the specificity of marketing for cultural organizations
- Distinquish between traditional marketing and marketing culture and the arts
- Introduce a marketing plan for the arts

INTRODUCTION

This chapter sets out some of the fundamental elements of the marketing model as applied to arts and culture. It provides a broad overview of each component of this marketing model. It also serves as an introduction to the discussion of the planning process, the central topic of this book.[1]

1.1 DEFINITION OF THE WORD "MARKETING"

The purpose of marketing is to optimize the interaction between a business and its clients and to maximize mutual satisfaction.

The concept of marketing consists of four main elements: the consumer's need, the satisfaction of that need, the connection between the business and the consumer, and the desire to achieve financial equilibrium (or, in the case of for-profit organizations, to optimize profits).

There are two ways of looking at marketing: in terms of satisfying market needs, and in terms of achieving the corporate mission. In the first case, the company tries to determine consumers' needs and give them exactly what they want, even if it means modifying its product or other business variables (including its mission, if necessary) until the customer is satisfied. Companies that operate this way are said to be "market oriented" and are profit-making enterprises. This type of market orientation also exists in the arts sector, in such forms as profit-based companies that offer a strictly "commercial" product like Broadway shows and Hollywood films. Cultural enterprises that are "product oriented" – usually not-for-profit organizations – take the artist's vision as their starting point and look for the consumer segment that is most likely to be interested in that vision. Companies in the leisure sector start from their mission statement and offer a product that corresponds to their objectives. Marketing must be seen as a tool that companies use to fulfil their mission. Marketing has to reflect the needs set out in the mission statement.

Marketing as it applies to product-oriented cultural enterprises can be defined as follows:

> The art of making contact with market segments that are likely to be interested in the product, by adapting the marketing variables – price, distribution and promotion – in order to put the product in contact with a sufficient number of consumers to achieve the objectives set out in the company's mission.[2]

Figures 1.1 and 1.2 illustrate the marketing process for a market-oriented company and a product-oriented company.

Figure 1.1
The Traditional Marketing Model

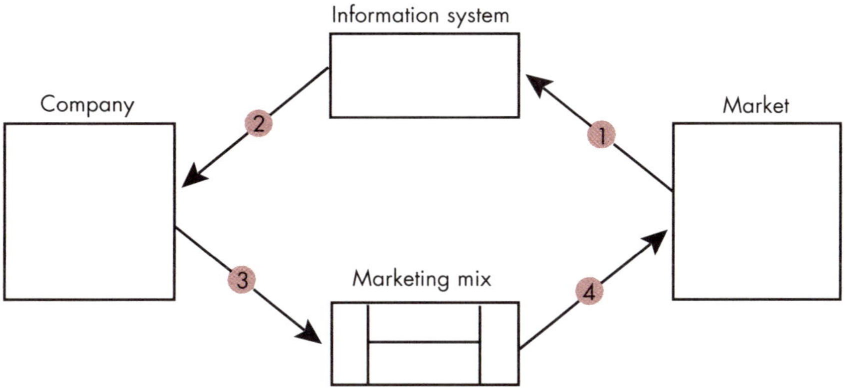

Figure 1.2
The Marketing Culture and the Arts Model

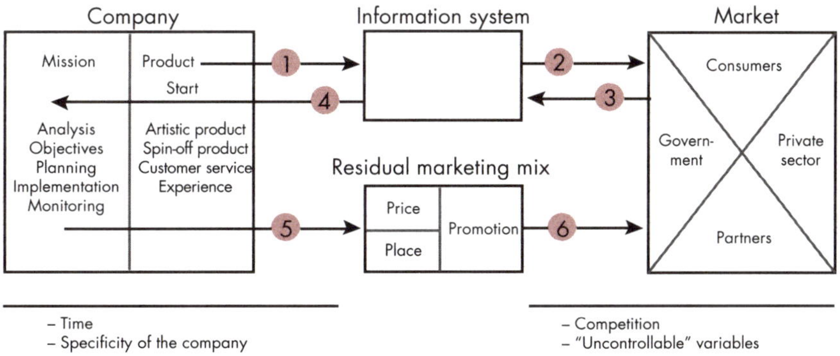

The traditional marketing process begins at the point where needs are defined by the information system. The company then decides whether or not to broach the market in question based on the resources at its disposal. For-profit companies found in the cultural sector are mainly market-oriented and can be said to use the traditional marketing model.

In the case of product-oriented cultural enterprises, the marketing process begins within the company itself, based on a work proposed by an artist or on the organization's mission. For arts enterprises, the mission and the artist's vision may be one and the same, particularly if the organization was created around an artist. For example, the mission of many dance companies founded by a choreographer is to perform that person's works. Often, in fact, the company is named for the choreographer. In the case of multi-product organizations (festivals, repertory theatres), although the company's mission is established by its founders, the enterprise is based on an artistic vision, not the creations of a single person.

All cultural organizations can be seen as positioned on a continuum, with product focus at one extreme and market focus at the other. Some are totally market-oriented, some are totally product-oriented and some lie somewhere between the two extremes.

1.2 WHY USE MARKETING?

The leisure sector has expanded remarkably over the past 40 years, especially in the area of culture. This enormous market growth has resulted in an impressive supply of products and services by a huge number of companies.

Consumers have a vast selection of options from which to choose. In large cities, it is impossible for one person to see or buy everything, even with an unlimited budget and unlimited time. Therefore, companies must clearly identify the type of consumer who is likely to be interested in their product and devise a strategy for ensuring that they receive a share of consumers' disposable income. In modern societies, cultural and product-based enterprises cannot be content to simply *offer* their products; they must also *sell* them. In other words, they must make sure they are understood and accepted by their natural or potential clientele and develop their customer base in order to ensure sufficient income – not only for the company itself but also for the artists it represents. This situation applies to all cultural markets.

1.3 THE MARKET

Cultural enterprises (and not-for-profit organizations in general) can target up to four different markets: the end consumer, governments, the private sector (sponsors and patrons) and various partners (distribution intermediaries, co-producers, distributors or specialized media). A specific marketing plan must be drawn up for each of the company's markets.

Each market usually consists of several segments, which can be defined as *consumer subgroups that share similar characteristics, express similar needs and desires, and respond to the same marketing strategy.*

Each segment requires an adjustment to one or more variables in the marketing mix. If it is not possible to adjust at least one variable, there is only one market segment, not two.

1.3.1 The Cultural Consumer

The consumer market is composed of individuals who buy a specific good or service.

Over the past 40 years, various surveys focusing on the sociodemographic profile of cultural consumers have been carried out in nearly every European country (both East and West) as well as in Canada, the United States, Australia and Japan. Whether conducted in the 1970s, 1980s, 1990s or 2000s, the surveys have all obtained the same attendance rates and the same sociodemographic profiles. They have consistently and systematically revealed strong polarization of audiences between high art and popular culture across all countries over the four decades. The surveys show, for example, that cultural products based on high art attract educated consumers, whereas those based on popular culture draw on all segments of the population, according to the relative weight of each. The proportion of university graduates making up audiences ranges between 50% and 70% for high art (symphony orchestras, arts festivals, art museums, etc.), compared with 10% to 25% for popular culture (pop music events, historical parks, etc.).

Other sociodemographic variables are also linked to attendance. These include average income (higher among consumers of high art than among consumers of popular culture) and type of occupation (white-collar workers account for a larger proportion of high art consumers while blue-collar workers are drawn to popular culture in greater numbers). It should be pointed out that this profile is based on averages. Less-educated individuals with lower incomes may be enthusiastic consumers of culture, as is the case for students and those working in the cultural milieu. Similarly, some people with very high salaries and advanced education are not interested in the arts and gladly keep their distance. Detailed analysis of the traits of the typical cultural consumer reveals other nuances based on the different disciplines. For example, dance audiences are younger and include more women than other performing arts audiences. Similarly, more women than men read novels, although a larger proportion of men read daily newspapers. In the film sector, there are two very different segments of avid cinemagoer, one dominated by young people (15–25 years of age) and the other by educated people; the majority of consumers in the film sector belong to one of these two segments. In terms of age, in the traditional high arts like opera, symphony concerts and ballet, most of the patrons are over 50, while new media attract a younger crowd.

1.3.2 The Partner Market

Although some companies sell their product directly to the end consumer, others use the services of a partner. A partner is an organization with which a firm associates on an ad hoc basis while sharing, to varying degrees, the risks and successes arising from an operation ultimately intended for the consumer market. Cultural enterprises have four main types of partner: distribution intermediaries, co-producers, distribution partners, and the media. Each can be considered a market in itself or a segment of the partner market.

In the performing arts sector, for example, a local promoter or presenter is a distribution intermediary. A touring company will turn to this local promoter to help it reach consumers living in a specific city or region.

Many cultural productions would never see the light of day if the producer was unable to find a co-producer to support the project financially. It is very common for a film to be co-produced by a foreign partner. This approach not only ensures adequate financing for the project, but also increases the film's chances of success as a result of the partner's familiarity with the local market.

A company may decide to join forces with another producer, or, alternatively, it may seek to form a partnership with a distributor. Distributors at both the local and the international level often play a similar role; some dance companies, for example, find foreign distribution partners when they accept a residency with a partner.

Finally, most cultural enterprises strive to secure the cooperation of the media in promoting their product. This is especially true of small companies that cannot afford ambitious advertising campaigns. In such cases, the media represent a key partner that the company must try to win over. A conscientious marketing manager will develop a strategy that takes the benefits perceived by this submarket into account.

1.3.3 Governments (the State) as a Market

The different levels of government – federal, provincial (regional) and municipal (local) – support cultural enterprises in various ways. Governments play a dominant role in the cultural sector in most Western countries. Sometimes they act as a consumer, or they may intervene to varying degrees, from acting as a simple partner, to playing the role of patron, to controlling the entire cultural sector of a nation.

The notion of the state as grant-giver and as a market implies that a cultural enterprise must define its own strategy to convince decision-makers to become partners in its activities. Each cultural enterprise faces competition from others in the same field, and its efforts to obtain more government support are essentially efforts to capture a larger share of a specific market. Since the government's arts budget is insufficient to meet the needs of the entire cultural sector, the arrival of a new company or the success of an existing one may mean that the money allotted to one company will be reallocated to the benefit of another.

1.3.4 The Private-Sector Market

For our purposes here, the private sector encompasses individual and corporate donors, foundations, and corporate sponsors.

Support can take two forms: contributions, and sponsored events or products. Contributions may be made by individuals, foundations or companies. Sponsorships come mainly from companies. A contribution is normally a philanthropic act, whereas sponsorship is a promotional initiative in exchange for publicity or advertising. Sponsorships are awarded based on promotional benefits that are calculated in advance. The corporate sponsor then evaluates the performance of its investment in terms of visibility, top-of-mind awareness and the vehicle's reach – that is, the number of consumers receiving the message.

1.4 ENVIRONMENT, COMPETITORS AND COMPETITIVE ADVANTAGE

A marketing strategy cannot be drawn up in a vacuum; there are many external restrictions affecting the market and the firm. The environment, including both the company and the market, is composed of two elements that constantly influence all organizations: macro-environmental variables, also known as "uncontrollable variables," and competition over which the company has some control.

1.4.1 The Environment

Macro-environmental, or uncontrollable, variables constantly affect the life of any corporation, which may have to adapt to radical changes yet never have the chance to act upon the causes of these changes. There are five main variables in the macro-environment: demographic, cultural, economic, political-legal and technological.

Demographics play a key role in the market, since a shift in the population may mean a rise or fall in demand. How the population is spread out within an area, which age groups dominate and which ethnic groups live there are only a few of the important dimensions of the environment that influence marketing. The values of a society, also called the cultural environment, play a key role in the marketing of a product. As values change, so do consumers' habits.

Firms, like individuals, must deal with their economic environment. Today, inflation, unemployment and recession are household words. During a recession, there are fewer potential consumers and fewer dollars available per consumer. This situation affects not only cultural enterprises but also sponsors: The corporate budget for sponsorships or donations shrinks immediately.

Laws and regulations are another key variable, since government action can radically change the face of an industry. A direct tax on the price of cultural products may lower demand. The effect of government intervention or action may also be positive – for instance, tax measures designed to stimulate the film industry.

Every company is subject to influences from the technological environment. Science has made tremendous advances and discoveries that have left few areas untouched. Technology has also had an impact on the arts. In some areas, such as audiovisual equipment, there is ongoing competition spurred on by innovations that constantly flood the market and can radically change the way a work is produced or distributed. Other areas may be less affected by technological development in terms of production and preservation yet be greatly affected in terms of distribution.

Two other elements must be considered in any marketing analysis: time, and the specificity of the firm. All companies must work within a changing environment. Market conditions evolve over time, as do economic agents, needs and tastes. Marketing should be considered an evolving process, and any strategy should be periodically reviewed and adjusted according to the environment and corporate priorities. Every organization has its own personality and acts as an individual entity. What may be an excellent marketing strategy for firm A could prove hopelessly inadequate for firm B.

1.4.2 The Competition

Competition is often defined as a semi-controllable variable. That is, even if a competitor's strategy cannot be directly affected, there are various ways to react – for example, by following the competitor's lead and lowering prices or by matching a lively advertising campaign with an equally lively promotional campaign. A firm is not as powerless when confronted with semi-controllable variables as it is when facing macro-environmental variables.

In each market segment, there is at least one competitor vying for the same clientele. To counter the competition and attract consumers, companies should have something in their marketing arsenal that is superior in the eyes of their clientele. In marketing terms, this superior element is called the "competitive edge" or Unique Selling Proposition (USP): something that is unique to a given company, something its competitors do not have or cannot get.

For example, some enterprises have an advantage in terms of product manufacturing. This may give them a very low production cost that their competitors simply cannot match. This low production cost allows the company to offer its product at the lowest price on the market. Some companies control the distribution market, while others find such a credible spokesperson that they can trump anything the competition might come up with. Others offer a product that is superior by virtue of its quality, unique nature or prestige quotient. Still others are known for their peerless customer service. In a highly competitive world, it is crucial that a company be able to set itself apart by giving consumers a reason why they should choose it over one of its competitors.

The cultural sector has an advantage over other sectors of the economy. Since artists generally strive to set themselves apart from their peers by carving out a creative niche for themselves, the competitive edge is simply the product offered by the company – the artist's work. Organizations that offer entire seasons (theatres and museums) or an extended program (festivals and community art centres) can gain a competitive edge by presenting artists or shows that are out of the ordinary. Companies that are unable, for any of a number of reasons, to distinguish themselves (in the eyes of the consumer) through their products must find an advantage in one of the other elements of the marketing mix.

1.5 SERVING THE AUDIENCE (QUALITY OF SERVICE)

Consumers have become extremely demanding with respect to service. And once obtained, their loyalty cannot be taken for granted; companies have to actively strive to retain their clientele. In highly competitive markets, customers who receive poor service may go elsewhere to find what they are looking for. This is certainly the case in the cultural sector. Why should customers continue to subscribe to a theatre that is not interested in giving them good service? There is always a temptation to switch to a competitor who offers an artistic product they consider to be equivalent.

Poor customer service can ruin the quality of the artistic experience, decrease the buyer's satisfaction and ultimately tarnish the company's brand image. The owner of a bookstore or the organizer of a cultural event who neglects this aspect risks losing his or her current or future customers. The role of customer service is to maximize customers' experiences with the company and manage their expectations of those experiences. Good customer service can set a company apart from the competition and enhance customer loyalty. It is much more expensive for a company to attract a new customer than to retain an existing one.

An efficient company that is interested in satisfying its customers will examine its "moments of truth." These moments occur every time a customer comes in contact with the company, whether by telephone, on the Internet, at the box office, at the cash, during an event or during the lodging of a complaint. The efficient company will analyze these moments of truth and seek ways to increase customer satisfaction at each point. It will "put itself in the consumer's shoes" in order to understand the consumer's expectations. It should be remembered that a customer's only contact with the company may be through its frontline employees: telephone operators, service representatives at the counter, box office clerks, ushers, waiters, guides and so on. It may be that this contact will determine whether or not the customer accepts the company's offer – whether he or she will repurchase. Careful selection and training of frontline employees and volunteers is not a luxury but a necessity.

Similar attention should be paid to the company's other markets. An ill-treated sponsor, for example, is unlikely to continue offering its support. However, even the best customer service in the world cannot compensate for programming that is of no interest to the client. That is why the product is at the centre of every marketing strategy.

1.6 THE MARKETING INFORMATION SYSTEM (MIS)

The marketing information system (MIS) is a fundamental part of the marketing process. It represents an arsenal of tools for enlightened decision-making. Of course, no tool ever invented can replace a manager's good judgement!

A MIS uses three types of data: internal, secondary and primary. Internal data represent all the information obtainable from within the company, such as sales and financial information. Usually, managers are not even aware of the abundance of useful information available within an organization. A good data-mining strategy can be of the utmost importance for marketing the company at very limited cost (see Chapter 2 for a full discussion). Secondary data are published by public or private organizations in the form of reports made available to the public in published form, via the Internet or through the library system. Primary data are obtained directly from the consumer. This information is usually collected by means of a poll, survey or market study or through business research. The company may carry out this work itself, or it may hire a specialized firm for the purpose.

1.7 THE MARKETING MIX

The term "marketing mix" encompasses four variables that we define as the firm's "controllable variables": product, price, distribution (place) and promotion.

1.7.1 The Product

Marketing defines the product as *the set of benefits as perceived by the consumer.*[3] This definition forces us to look at the product from the customer's point of view. When consumers buy a book or a CD, they are obtaining tangible merchandise; when they go to a movie, concert or exhibition, or when they participate in a municipal recreational activity, they are obtaining a service. But what they are really buying is not the actual product or service but the experience of consuming. This experience varies according to what the consumer wants – hence the interest in defining it as a benefit obtained by the consumer.

The importance of the customer's point of view cannot be overstated. Even in the case of a product-based approach, managers must always put themselves in the consumer's shoes in order to understand how consumers think, what their priorities are and what is important to them. This is the reasoning behind the idea that it is impossible to sell consumers something they do not want. A company may engage in false representation, but consumers rarely make the same mistake twice. The key to successful marketing is the ability to put oneself in the customer's shoes in order to understand and serve him or her better. This does not mean that the artistic vision should be modified, but it can be very useful in planning a promotional campaign or enhancing customer service.

By taking the consumer's point of view, the company will be better able to devise a strong positioning or repositioning strategy, build a solid brand and develop customer service that will set it apart from the competition; it will also be able to see that certain population groups will never be interested in the company unless it modifies its product, which is something that most not-for-profit enterprises, particularly cultural ones, have no interest in doing except in the event of repositioning by the artistic director.

Figure 1.3
The Product Life Cycle

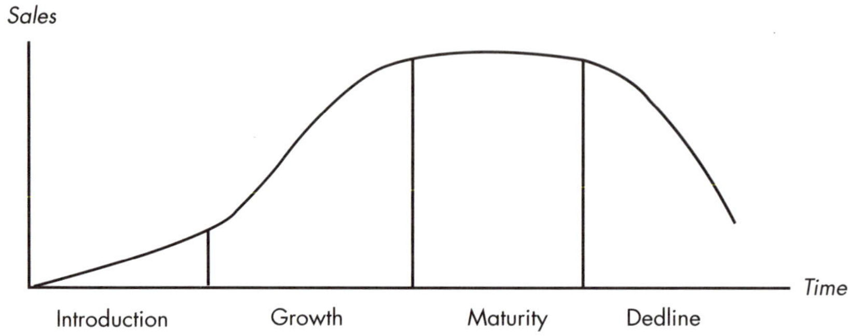

1.7.2 The Concept of Product Life Cycle

The concept of a life cycle for a product stems from the notion that everything, from people to products, is born, grows and dies. Some products know moments of glory, only to fall into disuse or oblivion. This is the case for papyrus, the quill and inkwell, and the gramophone. All of these products were replaced by more efficient products that fulfilled certain needs better than their predecessors did.

In short, the idea of a product life cycle exists because consumer needs and preferences change as technology evolves. Taste and technology are two interdependent phenomena that not only influence each other but often speed up a product's life cycle.

A product's life cycle consists of four stages: introduction, growth, maturity and decline. Although it is difficult to pinpoint the exact stage a product may be in at a given moment, each phase does have specific characteristics. Figure 1.3 shows the theoretical curve of the life cycle. This curve is actually the demand curve for a product over a period of time. The elegant line shown in Figure 1.3 is rarely reproduced in reality, where each phase may not be equal in length. Demand for a product may be a mere flash in the pan or it may blossom slowly and reach maturity years after the introduction stage. The number of possibilities is infinite: Some performing artists fade into obscurity after a single CD, while others, such as the Rolling Stones, seem to be able to remain in the maturity stage indefinitely.

The term "product life cycle" should be considered in its broadest sense. The concept may apply to a group of products offered on the market (market life cycle) or a specific brand or product. Generally, the life cycle of a market is composed of a series of superimposed product life cycles, which, in turn, are made up of superimposed brand life cycles. Organizations also have a life cycle. In the cultural sector, for example, a company that is closely linked to its founder or in which the founder and the organization are synonymous may come to an abrupt end if that person dies, retires or resigns.

1.8 PRICE

One defines a direction for the price variable by first answering the following question: What results do we want our pricing strategy to achieve? Naturally, every organization wants to maximize its income by determining *the price the consumer is prepared to pay.* To do so, however, it needs to choose among several types of pricing: preferred pricing (for gala events or benefits, for example), student pricing, last-minute pricing and discount pricing.

The price variable is not just a question of the amount of money needed to attend an event. The price of a product can be defined as a set of elements that include the monetary worth, the expense of going out, the consumer's effort (in researching and consuming the product or service) and the perceived risk of dissatisfaction.

Often, the cultural product is offered free of charge – or, to be more precise, at no financial cost to the consumer. As shown in the above definition, consumers must pay a certain price to attend the event, particularly in terms of the risk of disappointment and the potential waste of their precious leisure time. After all, the arts consumer cannot test the product before experiencing it. Customers must therefore always sacrifice something in order to attend an event, be it money, time, or both. There is always a price to pay, albeit not necessarily monetary. There really is no such thing as a free lunch.

Companies can obviously enhance the value perceived by the client through the use of an effective promotional strategy, product positioning or unique customer service. Consumers may then be more prepared to pay a higher price for the product – in terms of monetary cost, expenditure of time or the risk involved.

Setting a price for a product may be the most difficult decision a manager will have to make. We repeat: The fair price is *the price the consumer is prepared to pay.* Although that definition does nothing to help managers in their day-to-day work, it once again shows the importance of putting oneself in the consumer's shoes before making any marketing decision. Setting the price too high may discourage consumers, while setting it too low will deprive the company of precious income and, further, will send the market a message that the product being offered is of inferior quality.

1.9 DISTRIBUTION (PLACE)

The distribution variable refers to the delivery of the good, service or product to the consumer. Distribution may take the form of a tour for a performance company, a travelling exhibition for a museum, a venue for a festival or community organization, or the delivery of a film (in a movie theatre or on videocassette or DVD), book or CD. Delivery is effected through a series of distribution intermediaries. Organizations that deliver their product in a fixed location (such as a festival, bookstore, theatre or museum) and have no desire to move it elsewhere can use the easiest method of product distribution: direct distribution. Producers of a touring show or a travelling exhibition are obliged to do business with a distributor (an auditorium, another museum or an exhibition hall) and sometimes even to use the services of a booking agent to sell their show to distributors in the territory being targeted.

The distribution variable is usually broken down into three components: distribution channels or networks, physical distribution and commercial location.

Distribution channels or networks involve the choice and management of relationships with the intermediaries (e.g., agents and distributors) used by the producer to reach the consumer.

Physical distribution includes the logistics involved in transporting works of art or personnel and tour equipment to destinations or points of sale, or in transporting books from the printer to the bookstores.

Finally, the choice of location of the commercial venue, whether an auditorium, museum, festival or movie theatre, is strategically important. A shrewd choice can give a company a significant competitive edge. A bookstore whose clientele consists primarily of students would do well to locate near a campus. It would do even better to opt for the only location available in the area for its particular type of business, as its competitors would then be forced to locate further from the campus, and thus a greater distance from students. Libraries, theatres and museums, on the other hand, will greatly enhance their attractiveness by concentrating in the same area: Consumers are willing to travel greater distances in exchange for wider choice. This is the principle behind shopping malls, which offer all the products a shopper could desire under one roof, saving the consumer trips to multiple locations.

1.10 PROMOTION

The promotion variable is concerned with communication between a company and its customer (its four markets). This communication has four purposes:

- to inform customers of the existence of the organization and its programming and how they can access it
- to periodically remind consumers of the existence of the program and the organization
- to convince consumers of the superiority of the company's offer (positioning) over that of the competition
- to establish a relationship with the customer

There are four promotional tools available to get the company's message out to the public: paid advertising, personal selling, free publicity and sales promotion. And for each of these options there are many communication tools available: publicity and advertising (newspapers, magazines, radio, television), public relations, billboards, direct marketing, annual reports, intranet, Internet, derivative products, personal sales, employee uniforms and so on. Since any form of contact between a company and its market constitutes a means of communication, even the company's personnel cannot control what others will say about them.

For small not-for-profit enterprises with very limited financial resources, the two last tools in the list – free publicity and sales promotion – are the best ways to market their product. Companies try to attract media attention for their programming by arranging press conferences and public appearances by their artists. They sell promotional items, sponsor contests, reduce prices and provide demonstrations. They offer reimbursement, deferred payment, refunds and loyalty programs. They even dress their personnel in the company's colours. In short, they will do anything that entails little or no immediate financial outlay.

Since all the tools and means used serve the same purpose, it is vital that the company have an integrated vision of all aspects of its communication system. The company's press officer should be aware of any instructions given to its box office personnel, and the person who canvasses for sponsors should be up to date on the ins and outs of the company's media campaign. All forms of communication must be consistent, presenting the same image and pursuing the same goals. Nothing is more frustrating for a customer than to discover that an advertised offer is not on the company's Website or is news to the company's telephone personnel.

Nevertheless, even a cultural enterprise can avail itself of the most expensive promotional strategy – personal selling – without investing vast sums of money, simply by using a little imagination (see Chapter 3).

1.11 POSITIONING AND BRAND MANAGEMENT

Every company can be viewed as a "brand," even if its brand is nothing more than its name. The artists it represents and the shows it offers are what make the brand in the minds of its customers, ensuring its place on the market. This phenomenon is called "product/brand positioning."

1.11.1 Positioning and Repositioning

Since the notion of positioning or repositioning is central to every marketing strategy, the marketing plan must be based on the firm's strategic vision.

Positioning can be defined simply as *the place a product occupies in the consumer's mind as compared to products offered by the competition.* Repositioning is any modification of *the place occupied by the product* further to analysis of the situation. Managers must reflect on the kind of product they are offering, the market segment(s) they are targeting and the place their product occupies in the minds of consumers in those segments, as compared to the products offered by their competitors.

The positioning exercise forces the manager to think strategically: to look at the product, the market and the competition as three interrelated and interdependent elements. This type of analysis will produce such questions as: Where are we in the market? How is the market evolving? How big is the market? If we continue applying the same strategy, where will we end up in the market? How are we perceived in comparison with our competitors? What are our company's strengths? Is there something we do better than our competitors – do we have a competitive edge? Do we have a positive public image? Are our intentions clearly understood? If we had to come up with a slogan that would identify us immediately in the eyes of our clientele, what would it be?

This exercise should be carried out for each of the company's target markets. In fact, the positioning exercise is just as crucial for sponsors, governments and the company's various partners as it is for its customers.

Marketing managers must always remember that consumers base their judgements on their perceptions – regardless of whether those perceptions are valid. That is why it is so important to put oneself in the customer's shoes. What might seem clear to managers in terms of positioning or public message may be perceived quite another way by the people they are trying to convince.

1.11.2 Segment Positioning

Companies may wish to adapt their strategy to the needs of a single market segment. This type of strategy, called a *concentrated marketing strategy,* is used primarily by small organizations whose limited resources and very specific mission call for the targeting of a particular segment. In certain cases – a troupe specializing in theatre for young audiences, for example – this type of positioning is done for segments that are defined sociodemographically. In other cases, positioning is based on geographical considerations. In most cases, however, organizations position themselves according to segments that are defined based on what consumers want (benefits).

Positioning a product based on the benefits it offers and the segments it targets allows managers to determine which products are their main competitors and which ones are their potential allies.

Every positioning strategy is the result of careful analysis of the attributes and benefits sought by consumers or, for product-oriented firms, analysis of the segment interested in the benefits offered by the company. Without this analysis, positioning is just a theoretical exercise that will do little to support the company's strategies. It is therefore important to clearly understand the criteria on which consumers are basing their preferences and choices and to remember that customers are buying the "experience" offered by the company, not a "price" (even though price is a significant variable). While market studies are useful at this point, they are not indispensable; the manager's experience and thorough analysis of the market can provide equally sound results.

1.11.3 Competitive Positioning

Managers who have a clear understanding of the segments targeted by their product will be better able to position that product. In many cases, however, different products will appeal to the same segments and provide the same benefits. This is where competitive positioning, or "product differentiation," comes into play. In such cases, the company must offer its clientele an additional benefit to distinguish its product from those of its competitors. The image management of show business personalities is an example of this phenomenon. It involves the creation of an event that will set one "star" apart from all the others. By seeking to differentiate themselves, these personalities are looking for a competitive edge, something that potential clients will be unable to do without.

Finally, it is quite possible that, during the life of the company, the product will have to be modified or the company's mission will have to be altered.

This type of situation arises when the market changes as a result of a shift in consumer tastes and preferences, the arrival of new competition on the market, the death of the company's founder or the installation of a new artistic director necessitating a *repositioning* exercise.

1.11.4 Branding

Branding is an important facet of the marketing strategies of most commercial enterprises. Customers who are indifferent with regard to markets recognize the features of a product by the brand and use this to differentiate among products. A brand can be a name, a symbol, a logo or a drawing: The Disney Mickey Mouse ears, for example, constitute a brand that is recognized around the world.

All companies have a brand name, even if it is just the name of their business. The name of a well-known company evokes images in the minds of clients (consumers, donors, etc.) that they associate with a particular product. Even people who have never been to La Scala in Milan or to the Museum of Modern Art in New York (MoMA) have a pretty good idea of what they would find in those places.

The purpose of the brand, then, is to identify and differentiate the products of one company from those of another, or simply one company from its competitors. The brand represents the benefits offered by the company in a condensed form. It can constitute a response to consumer expectations. For example, when Céline Dion fans buy one of her CDs, they know exactly what to expect. Some people will also adopt a brand – of clothing, for instance – to consciously or subconsciously proclaim their membership in a particular group.

There are five characteristics associated with a brand: perceived quality, brand awareness (the name is well known), loyalty, association with salient elements, and association with identifiable tangible and intangible assets.[4] These characteristics can be defined as follows:

Perceived quality. The different markets and market segments can get a sense of the quality of a brand even if they have never used or consumed it. The definition of quality may differ according to the segment: The criteria of a knowledgeable classical music lover will be different from those of someone who is unfamiliar with classical music. Similarly, the quality of a festival will be defined differently by people who are simply seeking an event with atmosphere and those who purchase a ticket because they want to hear the artist being featured. It is important, though, to differentiate between quality as defined by experts (critics, other artists) and quality as defined by the general public.

Brand awareness. The larger the percentage of the population that is able to name a brand (with or without assistance), the stronger the brand.

Loyalty. Customer loyalty can be measured using the number of repeat purchases or the rate of subscription renewal.

Association with salient elements. For example, markets may identify the brand with the quality of a museum's collections or the work of an artistic director.

Association with tangible and intangible assets. The architectural features of the Sydney Opera House constitute an important element in the market's perception of its brand. This building is known around the world for its original design. The Sydney Opera House even serves as the logo for the Australian government's department of tourism.

The higher a brand scores on each of these characteristics, the stronger its image and the greater its market value. In the cultural field, a strong brand is particularly important since a cultural product constitutes a specialized purchase. Remember, we are talking not about *objective* strength/worth but about strength/worth as *perceived* by consumers. Even if experts view a brand as superior, if consumers do not view it as superior, the brand cannot be considered strong. Similarly, clients can have a false impression of the characteristics of a brand because, as we have seen, marketing has to do with clients' perceptions, which do not necessarily reflect objective reality. When this occurs, it is up to the company to launch a promotional campaign to clarify the situation.

1.12 THE MARKETING PLAN

The marketing plan of a profit-oriented company is based on the artistic product and the mission and on market needs, wants and desires in the for-profit sector. The plan relies on knowledge of the clientele and the company's target segments, sales objectives and customer service objectives. It comprises a positioning statement, a marketing-mix strategy and tactics, a contingency plan and a procedure for evaluating results.

1.12.1 The Marketing Plan Process

It is virtually impossible to draw up a marketing plan without going back and forth through its different steps. Marketing planning is first and foremost an iterative process that must be constantly evolving in order to reflect changes in the environment and in competitors' reactions. It is not unusual for a cultural enterprise to have to revise its marketing plan throughout the year based on interim results. A long-term plan usually has to be revised on an annual

Figure 1.4
The Marketing the Arts Planning Process

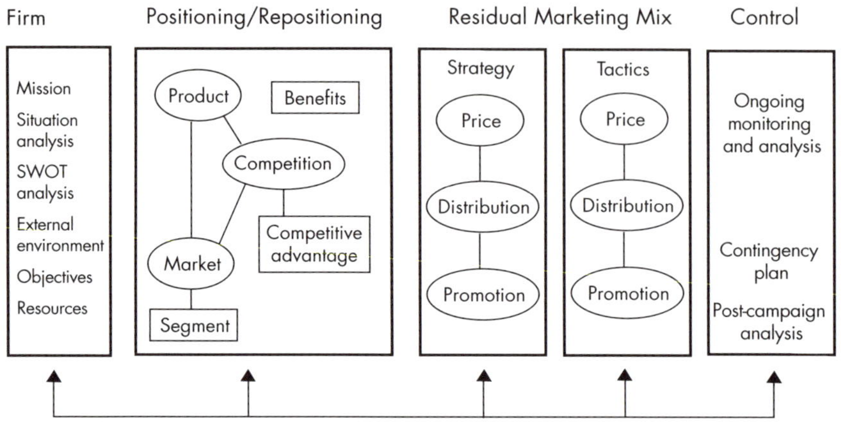

basis. A long-term plan is not intended to keep the manager in a straitjacket until the end of the period covered by the plan; it must be revised periodically to take into account the company's results and any new occurrences that could not have been foreseen when the plan was drawn up.

The marketing plan is an extremely useful tool. It gives companies a clear idea of the direction they should take and forces managers to look at all facets of each question in a structured and organized way.

Figure 1.4 illustrates the marketing planning process. As can be seen, this illustration of the planning process is just another way of representing the marketing model (see Figure 1.2).

1.12.2 Situation Analysis

It is essential to take stock of the company's current situation and to have a vision of where it will end up if nothing is changed. Based on that vision, a pertinent objective can be set. This exercise need not be exhaustive; it should be simple but enlightening. It is not necessary to expend great effort to arrive at a clear idea. In some cases the manager will need to gather data from the market; this can be done through data mining or market research.

Once you know where you are and where your current strategy will take the company if you do nothing to change it, it is time to set objectives. Although the marketing objective is based on the company's mission, it must also take into account the company's financial, human and technical resources.

Once a direction has been set (objectives) based on the company's requirements (mission, goals and resources), the manager is ready to tackle the three main steps in the marketing plan: positioning (or repositioning, as applicable), the other variables in the marketing mix and the monitoring of activities.

But first, a word about the terms *objectives* and *tactics.*

1.12.3 Objectives

Companies generally define their objectives over the short, medium and long terms. The definition of *term* varies according to the organization. One company's "short term" might be a year while another's might be just three months. Similarly, "long term" might mean ten years for one company and five for another. It is up to the manager to decide on the horizon that best suits the company's situation at a given point in time.

It is essential that a company set measurable objectives – measurable in terms of quantity rather than quality. "Increased sales" is not an objective that can easily be measured at the end of the period. It will be easier to determine whether the objective has been met if it is expressed as a figure: for example, "20% sales increase." This does not imply that companies should avoid setting qualitative objectives, however; it is important to be able to qualify what the company wants to achieve. We are simply stressing the importance of the operational aspects of any marketing action.

Once the global objective has been set, it can be broken down based on target markets or segments, or on the type of clientele the company wishes to attract.

1.12.4 Strategies and Tactics

There is a difference between a strategy and a tactic. A *strategy* originates in an overview of the means to be used to achieve a final objective (for example, hiring booking staff instead of using an agent to book internationally). A *tactic* is a specific adjustment to an element of the strategy at a given point in time (for example, attending an important trade show instead of sending costly brochures around the world and then making telephone contact). A manager can devise and implement several tactics to help achieve the objectives of the company's strategy without changing the strategy itself in any way. Sometimes we use the word "means" rather than "tactic," as it can amount to the same thing.

Once again, strategies and tactics can be determined for both the company as a whole and its marketing function. All marketing strategies and tactics, however, will be based on the company's overall strategy.

1.12.5 Planning the Marketing Mix

Each of the four marketing variables can be planned with respect to two main elements: the strategy to be followed, and the tactics pertaining to each strategy.

The strategy as it applies to the product variable in an arts context is linked to the vision of the artist or the artistic director; this vision will determine the path for all of the company's artistic activities. The exact art works to be offered to patrons will then be selected.

The positioning strategy, including the branding of the company and the level of service to be offered to the clientele, must then be determined. Based on the artistic vision, the marketing manager fixes the target segment, defining the benefits to be offered to potential patrons according to the company's competitive edge. This strategy with respect to the product leads to the different tactics to be employed.

The same model can be applied to the other three variables. A sense of direction is given to the price variable – a multi-segment approach, for example, will lead to a different price for each consumer segment (30% off for students, regular price at X amount, 50% last-minute discount and so on). The same sense of direction and tactics can be used for the distribution and promotion variables.

1.12.6 Monitoring the Effectiveness of the Marketing Program

Any marketing plan must consider the unforeseeable and be able to determine if the company's objectives have been reached. Managers need to ask themselves the following questions: How am I going to measure my progress and decide if I am still in line with my objectives? At what point should I start wondering if I'm headed for disaster? What can I do to stay informed? What methods should I use to adjust my aims?

This is where a good marketing information system, particularly an internal data system, becomes important.

Regardless of the method a company uses to monitor results, management must be prepared for the unforeseeable and – most important – decide in advance what steps it will take to re-launch a campaign that is not getting off the ground. The company may choose to set aside some funds for extra advertising, offer free tickets to certain target groups, or reduce its expenses. The method used is not as important as planning for the unexpected and being prepared for the worst.

In the next chapter we discuss in more detail the different elements of a marketing plan. When comparing Chapters 1 and 2, you will notice that a marketing plan can be tackled in different ways but that the end results will be identical.

1.13 CONCLUSION

In this chapter we have introduced some of the basic marketing concepts to be applied to any marketing plan. These include managing the product, the brand, customer service, and the distribution, price and promotion variables. The marketing process also relies on market analysis and a system for monitoring results.

SUMMARY

The marketing approach for culture and the arts mainly concerns enterprises in the arts sector. Certain firms in the cultural industry have a product-oriented approach. For others, the traditional approach is adequate. Because of the high degree of competition in the arts and leisure industry, it is important for managers to know how to use the different marketing concepts.

Overall, marketing must be seen as a set of activities designed to draw the company and the consumer together. The four components of marketing are: the market, the marketing information system, the company and the marketing mix.

The market consists of the consumer, the company's partners, governments and the private sector. The marketing mix covers four controllable variables: product, price, place (distribution) and promotion.

Any company, in order to survive in the marketplace and serve its artists, must be aware of its competitive edge and must develop quality service and a strong brand. Throughout the product's life cycle, the manager must develop good positioning or, if necessary, reposition the company.

Finally, the marketing model can be developed into a marketing plan based on the company's mission and objectives.

Notes

1. For a discussion of the elements of the marketing model as it applies to arts and culture, see F. Colbert et al., *Marketing Culture and the Arts,* 3rd ed. (Montreal: Carmelle and Rémi Marcoux Chair in Arts Management, 2007); www.gestiondesarts.com.
2. Ibid., p. 12.
3. Ibid., p. 30.
4. N. G. Caldwell, "The Emergence of Museum Brands" (*International Journal of Arts Management,* Vol. 2, No. 3, 2000), p. 28–34.

PLAN

Chapter 2

The Marketing Planning Process

By Dan J. Martin

OBJECTIVES

- Describe key marketing concepts
- Discuss common barriers to marketing planning
- Describe the Marketing Planning Process
- Explain how to build a marketing plan

INTRODUCTION

This chapter will focus on the planning process: how to approach and implement marketing programs based on the institution's primary mission, vision, values and objectives, within the context of the community in which the institution is situated.

2.1 KEY MARKETING CONCEPTS

Before engaging in the formal Marketing Planning Process, your organization must have consensus on a few key concepts.

2.1.1 Key Concept 1: Art/Culture Is a Commodity but the Challenges Are Complex

In the 1970s in the United States, as the number of not-for-profit arts institutions multiplied and funding for them increased at a similarly high rate, the impression among those of us who were jumping into the industry was that we were doing "God's work." We believed – and many of us still believe – that the arts are not a commodity to be purchased and consumed but a precious part of our lives, like good health or religious faith – something vital to our very existence – and that they are not an industry but a "cause." That was and still is true for perhaps 3% to 5% of the US population (the "culture vultures") while possibly another 10% or so believe that the arts contribute to their quality of life. Like it not, however, for the general arts audience cultural events *are* a commodity, a good that they "purchase" and "experience" rather than consume. Thus the arts are not a consumer good like bread, eggs or milk but a specialty good. Furthermore, arts and culture fall into the large leisure market, alongside other entertainment options such as restaurants, movies, sporting events and vacations. Lenwood Sloan, who danced with the Martha Graham company and served as a program officer for the National Endowment for the Arts, once said: "No one sits at home after work, opens the paper, and says 'Gee, honey, let's have a not-for-profit arts experience tonight.' No, instead it's 'let's go out'." His point is clear: We compete in the larger marketplace of leisure, relaxation and diversion. Whereas marketers of consumer goods simply need to establish brand dominance among other, similar goods that the consumer needs to sustain life, marketers of the arts must convince people to embrace leisure activities as a vital, life-sustaining "good" and then choose the arts as one of those options. Marketing the arts is a far more complex process than marketing bread or paper towels.

2.1.2 Key Concept 2: Relevance and Resonance: Customer Orientation

The messages sent by our marketing efforts will be effective only if the target market understands why the work being promoted is relevant to it. The targeted consumer must be able to appreciate the value in participating in the experience, be it a performance or an exhibition. The consumer may need context for the work, samples of the artistry or some introduction to the experience in order to make a decision to engage. Why this artist or this work? Why at this time? What is it about the work or the interpreters of the work that makes it important for her to participate? For those who are predisposed to attending a concert, it may be enough to simply include the name of a soloist in an advertisement or in a letter sent by direct mail, but for the majority of the potential audience, more – much more – will need to be done.

To appreciate the relevance of the messages being sent, the target market must be able to comprehend them; the messages must resonate with the recipient. This concept can be referred to as *market sensitivity.* Most cultural managers understand the concept and use it in programming their events. Most performing arts institutions, for example, would not program a serious or potentially disturbing work during the holiday season, and we always start the season (and usually end it as well) with something "big." More popular work will be presented in larger venues, while work appealing to a smaller market segment might be featured in more intimate venues. In order to keep a tight focus on relevance and resonance, perhaps we should use the descriptor *customer orientation* rather than market sensitivity.

Customer orientation within a product-driven institution is not easy to achieve or maintain. The very nature of our product-driven internal environment can be a force in restraining or limiting customer orientation. Many of us working in not-for-profit arts/culture enterprises began as artists ourselves. It was "all about us" – about the creative passions within us. We were encouraged to create work that inspired us, not work that would "sell." We are now part of institutions designed to create or facilitate the work of artists, and we often allow that deeply engrained "it's all about us" perspective to guide our decisions. This may be the right vision, the right strategy, when it comes to producing the work, but it distorts and often thwarts our best intentions in communicating the work to our audiences. The result is, more often than not, *representation* rather than *communication.* Representation results when the staff of an organization believes that its marketing and communications materials position the organization favourably within the community from the staff point of view. If the staff is not oriented to the target market – intellectually, aesthetically, emotionally and/or physically (geographically) – the message will fall on deaf ears. The message must be delivered to the market segment where it "lives," and it must "speak" in a language understood by that segment. The vehicle or the language used may seem alien to those of us who work within the institution; we may feel that it does not represent us well, and we may want to change the message to something we find more appropriate. We must overcome that urge to be representational; we must be vigilant in maintaining effective communication with our market segments.

This disconnect is evident in the experience of a regional theatre company. At a board meeting, the company's managers handed out index cards and asked their trustees to jot down the words they used in describing the company to their friends. Upon reviewing the results, the theatre's artistic and managing directors were surprised. Many of the trustees used the words "experimental," "avant-garde," "edgy" and "controversial" to describe a company that produced contemporary American drama that the directors and their peers would not have labelled experimental. Members of the community, however, did use those words to interpret and communicate the kind of work being presented: unusual plot, non-narrative dramatic structure, coarse language and nudity. Such adjectives were commonly understood descriptors among the market segments for that particular kind of theatrical experience.

2.1.3 Key Concept 3: Marketing Is Every Interaction between the Institution and Its Audiences

Every interaction between a patron and the institution serves to enhance or negate your marketing efforts. Your marketing may move a patron with grace and style to purchase a ticket to the theatre on Friday evening. She may arrive at the hall eager to experience a wonderful evening of modern dance, open to what is about to unfold before her. If, however, the driving directions were incomplete, there were long lines at the box office, an usher seated her in the wrong section, or the building was too hot or too cold, any sense of excitement or anticipation you may have created for the patron could have been deflated. She may have left the theatre after the performance – or even at the intermission – with a neutral or poor impression, and a great deal of additional marketing may be necessary to change that perception. This is not to suggest that marketing departments must control every aspect of the organization, but everyone connected with the institution should know that every aspect affects the patron's experience of the work.

2.1.4 Key Concept 4: Points of Entry

In *Waiting in the Wings*[1], Bradley Morison and Julie Gordon Dalgleish present a concept that has stood the test of time: Not everything we do is for every person or for every audience. We need to find the right *points of entry* (POE) for new audiences. We must match the right work with the right audience at the right time. Arts participation is not immediately addictive; interest in the arts grows and develops as the experiences become richer and more challenging. Furthermore, introducing someone to the arts in a way that triggers a negative response because of a functional, social, psychological or economic risk could do more harm than good. Solid marketing research will help to identify appropriate POEs for new audiences and for existing audiences ready to increase their participation.

2.1.5 Key Concept 5: The Complexity of the Purchasing Process: Individualized Communications

Many marketing professionals use the AIDA principle to produce change in a consumer: gain the patron's (A)ttention, create (I)nterest in the patron, provoke the patron to make a (D)ecision and stimulate the patron to take (A)ction. This is the process from the marketer's point of view. From the consumer's point of view, the process is more complex; it begins with ignorance of the product and moves up through awareness, knowledge and understanding to the point of conviction about the value of the product; the final steps in the Purchasing Process[2] are deciding to become involved and taking action to complete the transaction.

The ultimate goal in marketing is to move patrons from wherever they are in the purchasing process to the very last point: Action! Our various market segments will be scattered up and down this process, from the very bottom – ignorance – to very near the top – decision or action. This principle speaks volumes about the need for individualized communication with our target

Table 2.1
The Purchasing Process Based on the Consumer's Product Knowledge

Step 7 Action
Step 6 Decision
Step 5 Conviction
Step 4 Understanding
Step 3 Knowledge
Step 2 Awareness
Step 1 Ignorance

market segments – possibly even customized messages to individual patrons within those segments – based on who they are and where they are in the process. In other words, successful marketing requires a variety of carefully crafted messages sent to the targets at the right point in time. One poster, one piece of direct mail or one type of advertisement is not going to serve all of our needs. The marketing tools needed to take a prospective patron from "awareness" to "knowledge" or "understanding" will be very different from those needed to motivate a single-ticket buyer to "decide" to purchase a ticket for the current production.

2.1.6 Key Concept 6: Prioritize Sensibly

Beware of superficial cost/benefit analyses. When faced with the inevitable situation of possessing far more marketing tools than you can possibly implement, you will likely prioritize your marketing initiatives instinctively based on high impact and low cost (field A in the matrix shown in Figure 2.1). This is not necessarily an incorrect response, but it is critical that you also consider how various marketing tools interact with each other: their sequential and interactive nature, their synergistic effects and so forth. You may discover that activities falling into fields B and C (increasingly higher cost initiatives) and fields D and E (less effective initiatives) are vital to the recipe.

Figure 2.1
Cost/Benefit: Sensitivity Analyses

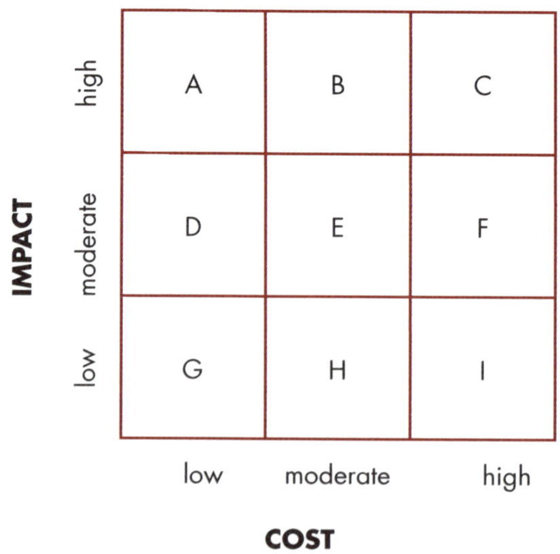

2.2 ADDRESSING BARRIERS OR OBSTACLES TO MARKETING PLANNING

There are a number of obstacles in the path to good marketing planning. They include:

- Lack of resources – time, energy, money and/or people
- Lack of support from the leadership
- The segregation of marketing from related programs and services
- Lack of proper orientation

2.2.1 Lack of Resources

Nearly all cultural enterprises – large or small – operate on limited resources. There never seems to be enough time, money or people to accomplish all of the tasks at hand or to take advantage of the opportunities, whether artistic or administrative, left at the doorstep of the organization. As a result, quick fixes and shortcuts become part of standard operating procedure. Quick fixes usually address the symptoms of a problem but not the root cause. Eventually a larger, more complete repair is required. In the aggregate, a quick fix can be more expensive than a comprehensive solution. Shortcuts are seductive because they may require fewer resources than a longer version of the same journey. Your shortcut may take less time, money and energy, but the longer route might have allowed you to economize by integrating the resources or experiences of others.

The return on the investment of human and financial resources in *good* marketing planning is always net positive for a cultural institution. The process will help to reveal actual and potential inefficiencies and redundancies as well as marketing efforts targeted at inappropriate or unreachable market segments. The planning process is not unlike the rehearsal process: It is an opportunity to reflect, review, practise and refine. A good planning process also unites the marketing staff – and others within the institution – behind a common set of strategies and objectives.

2.2.2 Lack of Support from the Leadership or Colleagues

Would a director or choreographer begin rehearsals with a clear vision of her ultimate goal but without a plan for how to reach that goal? Would she engage in a creative process without considering the capabilities of the artists involved or the resources available within the institution? Of course not! Why would the same not be expected of the institution's marketing director? The organization's leadership must understand that planning is as important to a marketing campaign as preparation and rehearsal are to a performance.

2.2.3 Isolation of Marketing from Related Programs and Services

As we have seen, every interaction between a patron and the institution impacts on marketing. All communications and external stakeholder outreach programs should be integrated into the planning process. The development (fundraising) department, for example, reaches out to almost the same constituency on a regular basis. Any department that provides some level of customer service – ticketing, reception, concessions, store, guards, docents, education – should be included in the Marketing Planning Process.

2.2.4 Lack of Proper Orientation

We have seen that *customer orientation* is a key concept in marketing. Good marketing planning focuses on the patron and on how best to communicate with her to effect any needed changes. We must embrace what good fundraisers refer to as the question not posed to a potential donor: What's in it for me? Good marketing reflects the same perspective: What's in it for the patron? In planning our communications, we must suppress our self-focus as artists and administrators and put the patron and her interests, needs and risks at the centre of the process. We need to show the consumer why and how she will be fulfilled or enriched by choosing our program.

2.3 THE MARKETING PLANNING PROCESS

The Marketing Planning Process is five-phase cyclical process:

- Centring
- Situation Analysis
- Strategy/Tactics Development
- Success Metrics
- Evaluation

2.3.1 Phase 1: Centring the Marketing Process around Mission, Vision and Values

Marketing, like all the other functions of a cultural institution, must serve the mission. This first phase need not be long, but it should be a formal part of every planning process. You should review the institution's mission, vision, and values and articulate what they mean for the marketing department. How can the marketing plan determine (albeit indirectly) and serve the mission of the institution? How will the marketing department's efforts help to guide the organization towards its long-term goals, its ultimate vision? How can the marketing department fulfil its obligations and remain true to the institution's core values? The marketing campaign must contribute to the audience's understanding of the mission and vision of the organization. While marketing must communicate with the target markets in language they understand, and in ways they understand, it must never misrepresent the mission or values of the institution. This is a fine line, and there may be disagreement among people committed to the institution as to its exact location. Careful consideration and articulation of the role of marketing in supporting the mission, vision and values will help to resolve any difference of opinion.

2.3.2 Phase 2: Situation Analysis

The Situation Analysis phase includes investigation of the internal and external environments, allies and competitors, and the conditions of the market(s) that the organization has penetrated or wishes to enter.

Internal Environments

Aesthetic/creative. What kind of work does the institution create or present? Is it considered high art or popular culture? Is it new, or part of a classical canon? Is the work presented in a polished style or is it developmental? Experimental or traditional? Edgy or comfortable? Professional or amateur?

Organizational/operational. How is the institution structured? How does it do what it does? How does the organizational structure impact on the creation or presentation of the work? Does this need to be reflected in the marketing in order for the target segments to understand the meaning of the work?

Resources. What are the human resources (paid and volunteer) and financial resources available to the marketing department, both immediate and long-term? The availability of resources over the long term will be useful in analyzing the sustainability and impact of large, multi-year marketing initiatives.

Capacity. There are two dimensions of capacity: audience capacity in terms of size, and programming capacity. Has the institution reached its limits in terms of audience size? How close is it to capacity? Regarding the capacity for new or expanded programs, can the institution create new programs or modify existing ones that might result in increased participation by existing and new market segments?

Current/past performance and trends. According to the participation data for the past three to five years, how many people are participating (reach) and to what extent (frequency)? What are the trends? Steady? Rising? Falling? Erratic? How does this period compare with the organization's previous periods?

Key institutional issues. Are there challenges or problems that may have affected the company's reputation or position in the community? How is the institution perceived by its current audience? Does the perception of the current audience match the institution's preferred or desired position?

External Environments.

You will have to analyze the institution's external environments in order to determine the institution's position on the regional landscape (literally and figuratively) relative to its target markets, its competitors and allies, and any other stakeholders. Before analyzing the individual environmental forces, determine the scale of the region. Is the region the city in which the organization exists? The metropolitan area? The state or province? The country? The global community? This regional scale will expand and contract from one environment to another.

Cultural. The environment is analyzed based on culture as defined by the level and kinds of arts and artists within the region, and culture as defined by the attitudes and behaviours prevalent in the community. What other cultural institutions are there in the community? Is the community rich in high culture? How diverse are its cultural offerings? How are these offerings received? Where does the institution fit in terms of the local, regional, national or international community? How does the community perceive "art" relative to other leisure activities? Is art valued? Is it considered old-fashioned, elitist or "boring"? Is it considered recreational, inspirational or, like exercise, something that is "good for you" but not always eagerly or willingly embraced? What is the general culture of the community: upscale, working-class, sports-crazed, class-conscious, open/friendly, closed/withdrawn? How might the cultural conditions exacerbate the *social risk* factor within the market segments?

Sociodemographic. Articulate the social and demographic characteristics of the population within the region using the traditional variables of age, gender, race, occupation, ethnicity and educational attainment. Pay close attention to educational attainment, as recent studies show that it is a strong predictor of arts participation. The current audience should be compared with the general population.

Economic. What are the general economic conditions of the region and the country? Can the community afford to support the institution at the level necessary to produce or present the work? With whom or what does the institution compete for resources? How might the economic conditions exacerbate the *economic risk* factor within the market segments?

Political. Is the current political climate favourable or unfavourable to culture? If it is favourable, how can this be leveraged by the institution? If unfavourable, how will this affect the *social* or *psychological risks* within the market segments?

Educational. Is education valued in the community? How many educational institutions are there in the region?

Technological. How ubiquitous is information and computer technology in the community? How comfortable are the people with that technology? What role does technology play in their decision-making about and consumption of leisure and the arts?

Geographic. Though not generally considered one of the traditional environments, geography does have an effect on cultural institutions. Is the institution close to or far from its market segments? What is the distance? How accessible do the market segments perceive the institution to be? Is the physical location welcoming, or do the target markets perceive threats or obstacles to participation?

Allies and Competitors

An important step in the Situation Analysis phase is identifying the institution's allies – other entities in the region that it might collaborate with or indirectly benefit from – and competitors – those it competes with for the same resources, attention and participation. A number of organizations or conditions might appear on both lists. Other cultural institutions, for example, increase the visibility and value of the arts in a region, and the well-known *multi-buyer* effect benefits all organizations, but patrons may have to make difficult choices based on their discretionary time and financial resources.

The Market

The next step in Situation Analysis is analyzing the market in which the institution exists. At this point in the planning process, you should identify the target market segments, both those the institution currently serves (and wishes to embrace more strongly) and those it hopes to engage for the first time. Formal market research by an external consultant can be undertaken if resources permit. A more common and less costly option is an in-house analysis of the market through a multi-stage exploration of the current audience of the institution, the broader audience for the cultural sector in the community, and how the balance of the population compares to those two subsets.

You can analyze the institution's current audience quickly and effectively using the sales database and patron surveys. A geographic information system (GIS) map of the addresses of subscribers/members and single-ticket buyers/single admission patrons can be compared with census reports to provide reasonably accurate portraits of the current audience. Audience surveys can help to confirm general findings and explore habits of particular interest to the institution (e.g., media used for information on arts programming, other arts interests). Surveys also can be helpful in more detailed stratification of the audience (e.g., specific characteristics of those attending on weeknights vs. weekends or afternoons vs. evenings).

Given the general understanding and acceptance of the multi-buyer effect, you can share your findings with your marketing peers at other institutions, in exchange for their findings. In some communities, the local arts service organization or cultural alliance will have completed market research across the region. In other communities, cultural organizations will have collaborated on "shared market research services." Regardless of how you obtain your information, determining how your audience compares to that of the cultural sector is critical to the Situation Analysis.

The third step in Market Analysis is comparing the characteristics of the arts audience to those of the general population of the region. Who else in the region "looks like" your audience? How many are there? Where do they live? How do they behave? What might be some of the obstacles to their participation? What other, competing, leisure or lifestyle activities might attract their attention and their leisure dollars?

You should also review the various consumer behaviour studies that have been published, including the time-tested VALS (SRI Consulting),[3] and the ongoing debate around the Family Life Cycle study by Wells and Gubar.[4]

At this point in the planning process, you may become discouraged. You most likely will discover that the regular audience for all of the cultural institutions in your region is no more than 10% of the total population. This is no reason to be alarmed; it is quite normal. Many of the recent studies and audience surveys that boast higher percentages of "participation" either report *total* attendance (ignoring frequency of attendance by individual patrons) or report attendance at any event that might be considered "cultural," including elementary school plays and toddlers' dance recitals.

The generally accepted level of participation in professional cultural programs has remained constant for nearly 20 years. Morison and Dalgleish's description of the breakdown of potential arts participants in *Waiting in the Wings* still holds true today. They place people into four categories: Yeses, Maybes, Noes and Ineligibles.[5] The Yeses (3% to 5% of the population) are those who are predisposed to arts participation because of their upbringing, education or other life influences. The Maybes (12% to 15% of the population) are those who, because of similar influences, might be interested in engaging in cultural experiences if the moment and motivations are in synchronicity. The Noes (50% to 55% of the population) are those who have made a conscious decision not to participate in the arts; culture will never be part of their lives. The Ineligibles (30% of the population) are those who, because of age, infirmity or incarceration, do not have an opportunity to participate in the arts.

If your institution has a market penetration of at least 3% of the general population, you may have secured many of the Yeses – those predisposed to participate – plus some of the Maybes. Market reach beyond 5% demonstrates successful penetration of the Maybes.

Additional market analysis should include investigation of the general trends in the marketplace, especially within the cultural sector. Is the cultural sector expanding or collapsing? Is leisure spending on the rise or in decline? Which communications channels – media, Web, wireless, print, telephone, direct mail – are the most active and effective in your community and for your market segments?

To complete this step in the process, you should identify the various target market segments for your institution. There are two common methods for identifying market segments: patron characteristics, and program characteristics. If you segment by patron characteristics, the variables may include the following:

- Age
- Family status/size
- Gender
- Educational attainment
- Geography (i.e., rural vs. urban, or neighbourhood status/personality)
- Lifestyle (attitudes, opinions)
- Cultural preference (type of programs preferred)

For segmenting by program characteristics, the following variables are common:

- Subscriber/member
- Single-ticket buyer/individual admission
- Lapsed subscriber/member
- Potential single-ticket buyer (e.g., patron of peer institutions)
- Cost-sensitive patron (e.g., student, senior citizen)
- Status-seeker

You could also segment the market across both categories of characteristics – for example, urban versus suburban versus rural subscribers. This level of segmentation could help you to focus your communications tools on the specific challenges involved in moving a market segment from one level of the *purchasing process* to another.

SWOT Analysis

The final step in Situation Analysis is synthesizing and summarizing the earlier steps in a SWOT analysis: an articulation of the institution's internal strengths and weaknesses and the external opportunities and threats as they relate to audience engagement and your efforts to facilitate the connection between the institution and its audience. A SWOT analysis should be done for each significant target market segment.

Strengths. The strengths should be relative to the competition – your basis for competitive advantage – addressing the unique or meaningful dimensions of your aesthetic, your mission, your capacity to serve, and/or your artists and staff.

Weaknesses. What aspects of the institution's operations or focus could be improved? Is the institution balancing efficiency and effectiveness well?

Opportunities. What artistic or programmatic "gaps" in the community could be filled by your institution? What existing goods, services or programs could be of benefit to your institution? What environmental conditions would serve to strengthen the marketing of the institution? Are there new markets that could be approached?

Threats. What external pressures are threatening the institution? What external conditions may be hindering the institution's ability to achieve its goals?

In the SWOT analysis, do not confuse internal and external points.

Once you have developed a good list for each category, sort the conditions or issues according to their importance to the institution. Prioritization is important for the next step in the Marketing Planning Process: developing strategies and objectives.

2.3.3 Phase 3: Strategy/Tactic Development

This is the moment in the Marketing Planning Process when you actually map out strategies and specific tactics to be implemented in meeting your annual marketing objectives; it is the operational centre of the marketing plan. "Strategies" are broad descriptive statements of the *direction* your marketing efforts will take in order to meet the marketing objectives; they will help you to identify the methods you will use to achieve your objectives.

Strategy Development

A strong Situation Analysis must be in place for an effective and focused set of marketing strategies and tools to emerge during this phase. Too often, stressed marketing professionals skip over the Situation Analysis or give it only scant attention and attempt to build an operational marketing plan based on "hunches" and "gut feelings." Such an approach ignores the critical foundation of the marketing plan.

Even when a strong Situation Analysis has been developed, some marketers stumble over how best to incorporate the many issues, influences and opportunities that have arisen during the second phase. They may cobble together some ideas based, once again, on hunches or a few interesting ideas that emerged during the Situation Analysis. There is, however, a logical next step that will facilitate the development of a more comprehensive marketing plan: a modified version of Heinz Weihrich's TOWS Matrix.[6]

The TOWS[7] Matrix is a planning tool frequently used to identify operating strategies based on the SWOT analysis. The process is quite simple and can yield valuable results.

The TOWS Matrix clarifies the relationships among the issues identified in the SWOT analysis, breaking them down into four categories:

- *Strength/opportunity strategies.* How can you leverage your internal strengths to capitalize on external opportunities?
- *Weakness/opportunity strategies.* What external opportunities can help you to reduce or eliminate an internal weakness?
- *Strength/threat strategies.* How can you leverage your internal strengths to reduce or eliminate your external threats?
- *Weakness/threat strategies.* What additional strategies could you employ to reduce other weaknesses and threats?

Table 2.2
TOWS Matrix

	Strengths	**Weaknesses**
Opportunities	**Strength/ opportunity strategies** Capitalize on internal strengths and external opportunities	**Weakness/ opportunity strategies** Take advantage of external opportunities to reduce or eliminate internal weaknesses
Threats	**Strength/threat strategies** Take advantage of internal strengths to reduce or eliminate external threats	**Weakness/threat strategies** Other strategies to reduce or eliminate internal weaknesses and external threats (primarily defensive strategies)

These strategies should be developed for each key market segment. Each strategy statement should be directional, not operational, focusing on a single approach. For each strategy statement, you should provide a rationale: the reason why this strategy matters and/or the market segment to be impacted. The rationale could describe the gap between the current and desired condition for that segment. After identifying the strategies for each segment, you may find it helpful to combine identical strategies. If properly developed and described, the strategies will lead to clear marketing objectives and tactics for each segment.

The following strategies are frequently used in arts/culture marketing:

- Advertising strategies
- Branding/positioning strategies
- Packaging strategies
- Pricing strategies
- Promotional strategies
- Publicity strategies
- Target market strategies

Advertising strategies. Provide a clear indication of how advertising will help you to meet your marketing objectives. Will your advertising focus on individual events, the full season of activity, the institutional brand/position, or some or all of these elements? Advertising designed to sell tickets – push the product – will look very different from advertising designed to set an institutional tone.

Branding/positioning strategies. If you are considering branding your institution, or creating a sub-brand of activities or services, these strategies will provide direction. Branding is an expensive process, and re-branding can be especially so. Branding and positioning strategies help audiences to understand why product-driven institutions do what they do and why it should matter to them. Build your image carefully and consistently, keeping in mind not only the competitive forces you face but also the perceptions of your audience.

Packaging strategies. Packaging strategies are not concerned with the "package" or container in the retail sense of word. Rather, they are directed at how you present your creative product to your patrons. Does the packaging of your events add to their value in the minds of potential patrons? Does it let your patrons experience the work easily and comfortably? Do the benefits bundled into the package attract, distract, enrich or repel? The purpose is not to modify the product itself but to explore ways of presenting it to audiences.

Pricing strategies. Cultural institutions use pricing strategies to achieve more than marketing objectives. One of the reasons for obtaining not-for-profit status and public subsidies is to make the work accessible to the largest possible audience. Pricing strategies are an invaluable part of the marketing process. On the psychological level, price infers value. Price also affects the brand or position of the work or the institution relative to the competition. And although many believe that demand for cultural products is inelastic, we know that pricing strategies such as discounts for specific times, seat locations or constituencies can be very effective in motivating consumer behaviour.

Promotional strategies. These are the foundation of your tactical plan. Keep in mind that you are not articulating specific tactics but providing direction for promotional activities. What behavioural changes are you hoping to achieve? Do you want patrons to linger longer in the gallery spaces? Museum visitors to come more often? Theatregoers to increase their attendance during the season? Patrons to speed up their purchasing decisions?

Publicity strategies. These provide direction for your media activities and related public relations efforts, including speaking engagements, publicity stunts, photo opportunities and story placement in various media. Describe how publicity will complement your other marketing strategies or how it will be focused on specific market segments.

Target market strategies. Target market strategies detail the weight you are placing on each market segment and the method(s) you will use to stimulate the appropriate change in those market segments. For example, do you want to motivate your single-ticket market segment to attend multiple events during the coming year? What approaches will you take to make increased participation more attractive to that segment?

Establishing Objectives. At this point in the process, you will establish specific marketing objectives for your target segments. You will base these objectives on the organization's programs, products and services as well as on your evaluation of the opportunities and challenges in each market segment and the location of that segment (and possibly sub-segments) in the purchasing process.

First, articulate marketing objectives (sales and revenue goals) for each individual program, product or service:

- season tickets or annual membership packages (full season/year and partial or multi-event offerings)
- individual performances or exhibitions
- programs or events outside of the traditional "season" offerings
- special services or amenities

For each program, product or service you have articulated, identify the specific target market segments for each offering, including sales and revenue goals for each segment. You may find it useful at this point to identify participation objectives for each segment:

- Audience Development: attempting to penetrate new market segments (increasing your reach into the market)
- Audience Maintenance: motivating existing segments to remain involved
- Audience Enrichment: providing additional services or benefits for a deeper, richer or more enjoyable patron experience
- Audience Expansion: encouraging patrons to engage with your organization more often

Obviously, some juggling of sales and revenue goals will be necessary if the market segment sales and revenue goals do not match the broader program, product or service goals.

Before analyzing the market segments for appropriate strategies, determine the location of the segments (and possibly the sub-segments) in the purchasing process. Do the urban empty-nesters appreciate your organization and its mission? Is the next step to persuade them to participate and take action? If so, they are very different from those suburban young families who have no knowledge of your organization. Your marketing strategies and tactics will be different for each of these market segments. The biggest mistake made by inexperienced marketers is to assume that "one size fits all" – that a single ad campaign or direct-mail package will reach a variety of market segments and motivate them to move from wherever they are in the purchasing process up to the "action" step. Even experienced marketers take this stance and justify their actions based on resource limitations. Certainly there are individual marketing tactics that serve multiple market-segment strategies and careful use of resources is paramount for all not-for-profits, but inappropriate marketing messages are wasteful.

The Marketing Mix: The Four Ps – Product, Place, Price and Promotion

At this stage in planning a campaign, most marketers begin structuring their tactics according to their strategies and objectives based on the traditional Four Ps: Product, Place, Price and Promotion. For more than a decade, marketing experts have been calling for a more customer-oriented approach to the mix. For instance, Lauterborn, Schultz and Tannenbaum suggest replacing the Four Ps with the "Four Cs."[8]

- Consumer value (vs. Product)
- Consumer convenience (vs. Place)
- Consumer cost (vs. Price)
- Consumer communication (vs. Promotion)

Consumer value. As we have seen, not-for-profit cultural enterprises are product-driven; you will rarely if ever find a resident artist or curator modifying a work based solely on what will "sell." Product is not one of the Four Ps that arts marketers have exploited to its full potential. If, however, you think not of the product but of its value to or relevance for the patron, you will discover opportunities to win her over to the organization. Your marketing should make the patron understand why she should see the work, why it might be meaningful for her at this point in time and how it might be of value to her.

Consumer convenience. Place is another tool in the marketing mix that is rarely available to the arts marketer. Place usually means "distribution": the physical location where a product is purchased or consumed. For most cultural organizations, the place or distribution point is fixed, whether it be concert hall, theatre, gallery or museum. If, however, you think of convenience of access, myriad tactical opportunities will emerge. How, when or where one can purchase a ticket may represent a convenience. A partial season-ticket package is a convenience. An early starting time for a concert is a convenience. The ability to reserve a time to view a popular exhibition is a convenience. Easy access to background information on a work or its creator is a convenience. Special amenities (e.g., parking, a members' lounge, priority seating) are conveniences.

Consumer cost. Pricing strategies have been used by arts marketers since the dawn of organized exhibition and performance. We use variable pricing and related strategies to draw cost-sensitive market segments to the work and to sell less desirable seats or make undesirable dates and times more palatable. Some managers use high prices with subsequent discounts to position the organization as prestigious (initial high prices) or to motivate market segments to take action (the subsequent discount offers). It is important, however, to look at the full "cost" of customer participation. What are the additional financial costs of attending an event? They could include parking, gas and tolls, meals and babysitting. What is the cost of purchasing a ticket in terms of time and effort? What are the potential social, psychological or emotional costs to the patron (the "risks of involvement")?

Consumer communication. Think beyond promotional tactics and gimmicks to the process of communicating with your audience and providing a level of service beyond the presentation of the work. Unlike the diner on the highway, we hope for more than a single visit from a tourist; we want to build a long-term relationship with our patrons. Although we use traditional promotional tools to attract our target market segments, we create a two-way relationship and then strengthen it over time. For cultural organizations, patron services – called "customer service" or "customer relations" in the private sector – is not a luxury and must go beyond traditional subscriber or member services. Hoteliers know that every hotel offers a bed, bathroom, television and phone. What positions one hotel above another is the quality of communication, service and interaction with staff. How well do you respond to complaints or problems? Do your ushers know what time the concert will end? Do your museum guards have basic information about the work being exhibited? Do your box office personnel know how to respond quickly to patrons with special needs?

Detailing the Tactics

The final step in this phase of the process is to develop the advertising, promotional and communications tactics you will exploit for each target market segment. Detail the specific media to be used, their intended target market segments, the message(s) they will convey, the strategies they will reflect, their expected outcomes, their timing (especially for sequential tactics dependent upon antecedent tactics), and the human and financial resources required to implement them.

2.3.4 Phase 4: Success Metrics

As George Thorn and Nello McDaniel of Arts Action Research have stated in many of their publications, if we do not establish metrics for gauging our success, someone else will – and those metrics may not be appropriate. Creating metrics for your marketing plan during this phase will not be terribly difficult if you have carefully followed the process so far. During the last phase, you identified sales and revenue goals and created a matrix of tactics for achieving them. The tactics have costs and timelines attached. During this phase, you established the three traditional categories for measuring the success of a marketing campaign: goals, budget and schedule. There is a fourth category for measuring success in not-for-profit cultural organizations: how well do the tactics support the mission, vision, values, and aesthetic of the organization and communicate them to the target markets?

The first three categories are easy to articulate. Organize your sales and revenue goals according to target market segments under each marketing objective. For each marketing tactic, list the implementation costs and draw up a timeline – the commencement and completion dates. As these are tangible and quantifiable metrics for success, you can expect little if any debate about whether you have met them when the results are being evaluated!

The fourth category – the degree to which your communication tools convey the mission of the organization or the degree to which your marketing reflects the values of the organization – is much harder to measure and analyze given the qualitative nature of the metric. Be sure to evaluate the effectiveness of your marketing messages in conveying the organization's mission and aesthetic from the patron's point of view – how the message is interpreted by the recipient – and not how well the internal stakeholders feel the marketing *represents* the organization to the community. Qualitative evaluation by a marketing consultant and/or carefully planned focus groups can provide valuable feedback for this metric. The degree to which the marketing reflects the values of the organization is easier to determine. Can a marketing campaign be deemed successful if it achieves sales and revenue goals based on misinformation or sensationalistic marketing techniques? (We think not!) Does the marketing over-promise or appeal to baser instincts?

2.3.5 Phase 5: Evaluation

The Marketing Planning Process comprises two levels of evaluation: ongoing monitoring and analysis during implementation, and post-campaign analysis and restructuring.

Ongoing Monitoring and Analysis

Many external factors will influence the impact and effectiveness of your marketing campaign. For example, the placement of your ads in the media and the delivery date of your direct marketing package can heighten or diminish the impact of your message. Rapid changes in the economy or even the weather can counteract the effects of the best marketing plan. Critical reaction to the work being performed or exhibited, especially by prominent media outlets, can make or break a marketing plan. Finally, nothing can guarantee the success of a work performed or exhibited; it is a fact of life in not-for-profit cultural organizations that not every creative venture will achieve its objectives. Implementation of the marketing plan must be continually monitored and evaluated. This is not to suggest that knee-jerk decision-making become the norm in the middle of a marketing campaign. It is simply a call for careful and strategic thinking and decision-making relative to any unanticipated external conditions.

Post-campaign Analysis

Strategies or tactics can be modified during a campaign. However, review, analysis and restructuring will have to wait until the campaign is over and research can be carried out. The effectiveness of various strategies may not be evident until all of the data are collected. Post-event qualitative research may be necessary to determine the degree to which people were motivated by the various communications tools. Just as some economists believe it is impossible to truly determine whether we are in a recession until it is nearly over, some marketing professionals believe it is impossible to truly measure the impact of a marketing campaign until it is finished, because of the interdependence of many tactics and the sequential nature of others. Post-campaign analysis will provide context for subsequent campaigns.

SUMMARY

There are no recipes for effective marketing, nor are there magic formulas for success in generating large, enthusiastic and engaged audiences. Successful marketing effectively and efficiently connects target markets with the work being presented or exhibited. Therefore the process is highly specialized, with specific messages directed to focused *market segments* based on the location of those segments – and sometimes even individual consumers within them – in the *purchasing process*. Marketing techniques cannot be copied wholesale from one program or service to another (even within the same organization) or from one institution to another.

Key marketing concepts

- Arts and culture is a product-driven industry
- Arts and culture is a commodity, but the challenges are complex
- Marketing provides relevance and resonance through customer orientation
- Marketing is every interaction between the institution and its audiences
- Points of entry (POE) – the right work for the right audience at the right time – is critical in bringing new audiences to an organization
- The complexity of the purchasing process calls for individualized communication with market segments
- Marketing tools must be prioritized sensibly and strategically

Obstacles to effective marketing planning

- Lack of resources – time, energy, money and/or people
- Lack of planning support from the leadership
- Segregation of marketing planning from related programs and services
- Lack of proper orientation

The Marketing Planning Process

- Centring the marketing process around the organization's mission, vision and values
- Researching the internal and external environments, allies and competitors, and the conditions of the market(s) that the organization has penetrated or wishes to enter
- Mapping out marketing strategies and specific tactics to be implemented to meet the annual marketing objectives
- Establishing metrics for gauging the success of the campaign
- Ongoing monitoring and analysis during implementation and post-campaign analysis and restructuring

Notes

1. B.G. Morison and J. Gordon Dalgleish, *Waiting in the Wings* (Washington: Americans for the Arts, 1992).
2. F. Colbert et al., *Marketing Culture and the Arts* (Montreal: Gaëtan Morin, 1994), p. 183–184.
3. *VALS™ Survey*: http://www.sric-bi.com/VALS/presurvey.shtml.
4. W.D. Wells and G. Gubar, "Life Cycle in Marketing Research" (*Journal of Marketing Research*, Vol. 3, 1966), p. 355–363.
5. Morison and Dalgleish, *Waiting in the Wings*, op. cit., p. 47–48.
6. H. Weihrich, "The TOWS Matrix: A Tool for Situational Analysis" (*Long Range Planning*, Vol. 15, No. 2, 1982), p. 61.
7. "TOWS" is "SWOT" in reverse.
8. R. Lauterborn, D. Schultz and S. Tannenbaum, *Integrated Marketing Communication: Pulling It Together and Making It Work* (Lincolnwood, IL: NTC Business Books, 1993), p. 12–13.

PLAN

Chapter 3

Integrating Information Technology into the Marketing Process

By Philippe Ravanas

OBJECTIVES

- Explore the impact of information technology on the marketing function
- Highlight e-marketing and online distribution techniques
- Present the concepts of customer identification and customer modelling
- Pinpoint the evolution of sales forecasting
- Apply dynamic pricing and yield management principles to cultural products
- Understand Relationship Marketing

This chapter sets out some of the fundamental elements of the marketing model as applied to arts and culture. It provides a broad overview of each component of this marketing model. It also serves as an introduction to the discussion of the planning process, the central topic of this book.

Over the last decade, the information technology revolution (particularly the combination of the Internet, database systems and computer analytics) has transformed the marketing function for the arts, bringing a level of sophistication previously reserved for bigger – and richer – industries.

Computers have brought about three "awesome powers."[1]

- *The power to record*: Companies can now store millions of customer records, with many characteristics for each – not just names and addresses, but age, gender, marital status and family configuration, and buying habits and history.
- *The power to find:* Individuals can be selected from a database by one or multiple identifying characteristics.
- *The power to compare:* Information on customers with one set of characteristics can be compared to customer information using a different set of characteristics.

Combining these three powers with the speed, interactivity and affordability of Internet communication sets the ground for a marketing revolution. It leaves few, if any, aspects of the marketing mix untouched, transforming the very notion of artistic product and its pricing, promotion and distribution strategies. For companies it brings the promise of more efficiency and for customers the promise of better service.

This revolution has just begun.

According to a famous forecast made by Gordon Moore, co-founder of Intel (the world's largest microprocessor company), the processing power and storage capacity of computers double every 18 months. This prediction was made 40 years ago and has held true since then. If it still holds true over the next decade, this could mean up to a hundredfold increase in capacity – enough to meet any conceivable need for computation. At the same time, the price of computers has been falling steadily. Furthermore, mobile phone technology offers the prospect of new, unchartered marketing territories.

This chapter will address some of the most salient innovations brought by the integration of information technology into the marketing process.

3.1 E-MARKETING

E-marketing is the application of the Internet to direct marketing. Direct marketing, which consists of direct connections with carefully targeted individual consumers both to obtain an immediate response and to cultivate lasting customer relationships,[2] is not new: Catalogue sales have been around for a long time. However, it has been rejuvenated by e-marketing. The Internet makes it possible for an organization to weave a dense, customized relationship with each individual customer, with the goal of better meeting consumer needs.

3.1.1 Direct Marketing Principles

In a direct marketing campaign, an organization sends a specific, usually time-limited, promotional offer directly to individual customers via mail, telephone or e-mail as opposed to via mass media such as billboards, press, radio and television. This generates a direct and rapid response in the form of an order, a subscription renewal, a request for further information, or a visit to a retail outlet or to a Website.

Mass media are effective for promoting fast-moving consumer goods but have proven less efficient for cultural products. In the field of culture, targets are usually too small, products too complex and budgets too limited for organizations to take full advantage of mass communication techniques. Moreover, the impact of advertising is undeniable but difficult to precisely measure. One can never be sure that an advertisement has been seen, understood and remembered, since communication goes one way: from the organization to its target. Separating the effect of advertising from other factors influencing consumers is particularly arduous.

Finally, promotional spending for consumer goods has been growing steadily and tends to overpower products with smaller budgets. Because of this constant growth, the multiplication of media and the resulting advertising omnipresence, consumers tend to become immunized to this type of marketing and shield themselves from the communication avalanche by ignoring most advertising messages.

Direct marketing can address all of these concerns. It offers several advantages:

- More accurate targeting, since only those consumers whose buying potential has been identified will be contacted with a customized offer. With traditional advertising, targeting is only as precise as the audience of a given medium. For example, an advertisement in a newspaper will be seen – and perhaps read – by the newspaper's readers, who are not all current or potential customers of the product being advertised.
- Accurate return-on-investment measurement of a promotional campaign, since the rate of response to the offer and the cost per contact are known. For example, if 1,000 letters promoting a specific offer are sent out and 20 people respond, the return on investment can be calculated by dividing the revenue from the 20 sales by the total cost (time spent

framing the offer and drafting the message and costs for paper, printing, envelopes, and postage and handling).

– A more personalized, two-way communication stream with the target, since the latter is encouraged to respond directly. This consumer feedback is essential for improving the service offered and customer satisfaction.

– The ability to send dense messages and complex offers without the space limitations of traditional media.

– Reduced lead time between the initiation of the offer and its reception by the customer, since the offer can be sent at any time, independent of formal publication schedules.

For all of these reasons, direct marketing is becoming the main means of communication for many cultural organizations.

3.1.2 Advantages of E-Marketing

The media traditionally used for direct marketing (mail, telephone) are still very much in use, but present some limitations:

Mail. Mail offers many formats for direct communication: letters, postcards, brochures, catalogues, samples, tapes or CDs. In many countries, the postal service offers discounted prices for mass mailing. It is nonetheless a relatively expensive and slow means of communication.

Telephone. Telemarketing (i.e., using the telephone for direct marketing purposes) is faster and more interactive than mail, but it is also more expensive. Calling people is time-consuming and labour-intensive. Telemarketing is regulated in some countries: Individuals can have their phone number "delisted" to avoid receiving sales and marketing calls.

E-marketing is a much more attractive and cost-efficient way of reaching customers. Access to the Internet is growing rapidly. In fact, the vast majority of cultural consumers are already connected. Sending an e-mail costs a fraction of the price of making a phone call or purchasing a stamp. E-marketing is as interactive as telemarketing but is much less intrusive, as the customer does not have to answer an e-mail message immediately. The lead time needed to launch a campaign is very short: It takes only a couple of hours to draft a message, extract a list of e-mail addresses and send an e-mail – against two to three weeks to draft, format, print and expedite a letter. As answering e-mail is easier and faster than answering a letter (no stamp or posting), the response rate tends to be significantly higher for e-marketing campaigns. Moreover, an e-mail can have multiple lives: It is easy to forward it to friends, thus transforming customers into advocates for the company.

For all of these reasons, the return on investment is vastly superior for an e-marketing campaign than for a mailing or telemarketing campaign. However, the flood of "spam" (unauthorized commercial e-mails) has tainted the image of e-marketing. In reaction, several countries have adopted restrictive legislation to curb the flow of unwanted e-mails. It is thus preferable to ask each customer for permission to send promotional e-mails, and to promptly remove their e-mail address from the database if they refuse. Also, the company should avoid sending each customer more than one or two promotional e-mails per month.

3.1.3 Application: Steppenwolf Theatre Company (Chicago, Illinois, United States)

In the 30 years since its founding, Steppenwolf[3] has become a beacon of theatrical excellence, receiving unprecedented national and international recognition. It has also grown from wild beginnings into a marketing powerhouse – without losing an iota of its artistic relevance.

Traditionally, Steppenwolf has relied heavily on its subscription base to sell the 200,000 tickets it makes available each year. The cost of attracting and retaining subscribers (season-ticket holders) is far lower than the cost of attracting single-ticket buyers, but, like many theatres in the United States, Steppenwolf has experienced an erosion of its subscription base.

Amy Singer Kaissar, a subscription consultant, explains: "Across the US, a seismic shift in ticket buying trends is underway. The subscriber, the member, and the season ticket holder are not renewing. They may still be coming, but they are not purchasing tickets up front and in bulk. This shift, which began in the late 1990s and continues today at an ever increasing rate, cannot be ignored. This sea change affects theatres, museums, symphonies, opera companies, trade organizations, sports teams and more. For theatres, long dependent on a guaranteed audience who pays up front, the waning is critical."[4]

Steppenwolf will thus have to rely more and more on single-ticket purchases – a costly proposition. "Selling single tickets can cost you 50 cents on the dollar in advertising," complains Executive Director David Hawkanson.[5] "If you take a page in the *Tribune,* the dominant newspaper in Chicago, you pay to contact its 1.8 million readers, but less than 4% of them actually go to theatre."

Whereas subscribers buy an entire season before a single performance has been presented (which guarantees audience levels and builds an organization's cash flow), single-ticket buyers are unpredictable – tending to buy at the last minute. With traditional means of communication, an organization lacked the time and the money to react to poor attendance. But because of the Internet marketing single tickets is faster and cheaper than ever before.

This is why Steppenwolf decided to invest in e-marketing, and to collect e-mail addresses by asking every ticket buyer for an e-mail address. "We place inserts in our program saying, 'Win a dinner at the restaurant next door by filling in this questionnaire'," explains Hawkanson. "The prospect of getting a free meal gets us a lot of information on our customers!" Steppenwolf has already gathered a database of 50,000 e-mail addresses. This database is an excellent tool for promoting ongoing or forthcoming shows, and it allows for quick corrective action in the case of lower-than-anticipated attendance. It takes only a couple of hours for the theatre to define a promotional offer, draft a message, source the database for prospects and e-mail the offer at very low cost.

Steppenwolf uses e-marketing not only to sell tickets but also to strengthen its relationship with its audience. "E-marketing is the easiest and most effective way to inform our audience," says Hawkanson. "If you pick up the paper today, you will read one review. We will give you six reviews to read in our e-newsletter. Feeding the intellectual curiosity of our audience is immensely important to us. We also use e-mails to build loyalty among our subscribers, particularly first-year subscribers, who always have the lowest rate of subscription renewal. We can track their subscription usage rate. We can e-mail them a reminder a day before each performance they have bought. If they don't attend one of the performances, we can send them an e-mail the next day and tell them, 'We didn't see you last night – anything we can do?' They need to be reminded that they have subscribed and that they can take advantage of a series of benefits, because they tend to forget. If they don't use their subscription, they won't renew it."

E-marketing is also a great tool for turning single-ticket buyers into subscribers. "The best way to have a single-ticket buyer come back to the theatre is to send him a promotional offer for another play as soon after the performance he attended as possible," explains Marketing Director Erin West.[6] "That's where e-marketing comes in. The day after you come to Steppenwolf, I can send you an e-mail saying, 'We are so glad you could come and see our show. Here is a two-for-one ticket offer for any other show you would like to come see over the next three months. We hope to see you again soon.' You could not have done this even five years ago, because getting the information about these customers out of the system [and] designing, printing and mailing a promotional letter was so laborious and time-consuming."

"We have also launched a blog,[7]" concludes Hawkanson. "We need this new outlet to share what our guests think about our new plays. We are also looking at ways to use phone text messages to keep our customers informed. All this will deepen our interactive relation with our audience. This is where we are going."

3.2 ONLINE DISTRIBUTION

3.2.1 Principles

Direct marketing often leads to direct distribution. Since the consumer responding to the offer orders the product from the company directly, the company can do away with intermediaries and resellers. This model best fits cultural organizations that host live audiences (theatres, symphonies, museums), because tickets can be easily mailed or printed from the Internet.

Online distribution offers clear advantages over traditional, "bricks and mortar" retail operations:

It is faster. The transaction takes place in real time, without delay or call back.

It is cheaper. There is no intermediary fee, no printing or mailing expenses, and no data-entry costs, since patrons enter their information themselves when they purchase their tickets.

It is better. Customers can buy their tickets at any time of the day or night, from anywhere in the world, and most of the information needed to complete the transaction is only a click away.

It is also more reliable: Selling online allows for better and more detailed sales reporting and analysis and for increased information on the profitability of marketing initiatives.

Inventory computerization has also allowed for tighter control – which has always been a challenge. "We have a huge number of unique products on sale at any given moment," comments Alan Levine, Chief Information Officer at the John F. Kennedy Center for the Performing Arts in Washington.[8] "Every single seat for every single performance is a unique item. So at the Kennedy Center we have as many as a million and a half unique items on sale at any given time. And we can only sell that seat once. We have to know what seat is next to it, because most people come to the show with someone else and they want to sit next to that person. So you have to be able to pull up a thousand of those seats and show their relationship to each other. And it all has to be done in real time. Subscriptions add a layer of complication. But ticketing follows strict rules: You can only sell a seat once; if you sell two seats they should be next to each other... And in software, if you have rules, you can program them."

An online distribution operation should be fully integrated and connected with the traditional, retail channels the organization might retain. That combination of online and traditional distribution is often referred as "click and mortar" distribution.

The Internet has also proven to be a very effective means of distributing music, images and film, although fostering piracy for these products. When it comes to selling non-digital products to consumers, the big challenge of "last mile logistics" arises. The last mile is the link between an online ordering process and physical product delivery. In contrast to store-based retailing, online shops have to deliver the product to consumers' homes, not just to stores. Last mile logistics has turned out to be a crucial factor in e-commerce. Severe delivery problems can occur, especially during periods of high demand such as holiday seasons. A customer who is disappointed because of poor delivery service is unlikely to shop online again.[9]

3.2.2 The Importance of a Good Website

A good Website is essential for online distribution. Yet many sites are substandard. They are unclear or even unreadable. Their navigation is counterintuitive and inconsistent. They lack basic product, price and availability information. All of this leads shoppers to become frustrated and fed up, never to return. It is also a waste of money. Developing a Website does not come cheap, and a bad one can damage the reputation of an organization.

For Eugene Carr, CEO of Patron Technology, "a great arts Website is a carefully structured collection of Web pages designed and organized to achieve a measurable goal, one that supports your organization's overall mission as well as your bottom line. A great arts Website offers visitors the information they want in the fastest way, with the fewest number of mouse clicks and the least amount of confusion. Such a Website is essentially a machine for marketing."[10]

The goals of a Website should be clearly prioritized. The Website should primarily be a tool for selling products, sharing product information and collecting e-mail. It should not be a showcase for creative and technical prowess. Therefore it should be controlled by the marketing department, not by the information technology department. A careful analysis of the site's visits should provide important information leading to improvements.

Particular attention should be given to the layout of the site. Navigating it should be logical, simple and intuitive. Unnecessary clicks should be eliminated. For instance, it should take no more than four clicks to complete the checkout process, and no more than two for express checkout. No order should be taken if the inventory is unavailable, and there should be no price surprises during checkout. In the case of physical goods, the site should have comprehensive order-tracking capability and offer precise delivery times. The site should be updated frequently, and inquiries posted on the site should be answered in a timely manner.

"Don't make me think!" exclaims Steve Krug, a renowned Website consultant. "It's the ultimate tie breaker when deciding whether something works or doesn't in a Web design. It means that when I look at a Web page, it should be self evident. Obvious. Self explanatory. If you have room in your head for only one usability rule, make this the one. Actually, there is a close contender: get rid of half the words on each page, then get rid of half of what's left."[11]

The development of a Website should be guided so that it will naturally attract visitors by earning top ranking on the major search engines (such as Google and Yahoo!) for selected keywords. The company should understand how people are searching for its products. Visiting the sites that are earning top ranking might give clues about how to improve the company's site. To optimize the chances of its site being picked up during a customer's Web search, a company might start the text of each Web page with relevant keywords, use these keywords as the links to the page and link each page with other relevant Websites.

3.2.3 Application: Sydney Opera House (Sydney, New South Wales, Australia)

The Sydney Opera House is Australia's leading performing arts centre, producing an eclectic mix of artistic and cultural activities and staging 2,400 events annually. Its magnificent site, its iconic shell-shaped building and its 3,000-seat concert hall attract more than four million visitors per year.

The Opera sought ways to expand the online experience of its patrons by enabling them to use the Internet to check availability, select seats and buy tickets in real time. To offer patrons the best service possible through their ticketing transactions, the Opera upgraded to a cutting-age ticketing system with a Web interface.

The result was clear: Online ticket sales increased significantly and now account for over a third (sometimes even 80%) of all single-ticket sales.

"Our online box office initiative has transformed the way we sell," says Claire Swaffield, Director of Information Systems for the Opera.[12] "First, it allows us to handle a much larger sales volume in a given amount of time. We sold approximately 60% of concerts for the Scottish comedian Billy Connolly online. We were able to support multiple transactions so that three concerts sold out within the first on-sale day. No way could we have handled that before. It's terrific to watch the Web channel ticking over and users coming through and being able to transact really, really quickly. We also can run our online box office for as many hours a day as we want, allowing our customers to buy tickets at any time of the day or night. Last, we now can capture, store and extract much more customer data than with manual transactions. We have a tremendous capability to market directly to our audience. More customers wanting to make purchases via this sales channel can now do so with amazing reliability. We have invested a lot of time and energy in our online distribution capability, but we are blown away by some of the things that we can do with it."

3.3 CUSTOMER IDENTIFICATION

3.3.1 Principles

John Wanamaker, the father of modern advertising, said: "Half the money I spend on advertising is wasted. The trouble is, I don't know which half." To avoid such waste, and to best prepare for an e-marketing campaign, an organization must first identify among its pool of existing or prospective clients those most likely to respond favourably to an offer. Today, technology enables enterprises to identify each customer individually and to use this information to approach him with specific offers.

The organization should collect all available and accessible information about each customer: contact information (name, address, phone number and e-mail address), data needed to assess his interest in the product being promoted (demographics, psychographics, consumer behaviour, satisfaction level, transaction details, etc.) and "clickographics" (Website-visiting behaviour and online transaction history). Some data, such as birth date, are stable and will need no alteration. Other data, such as satisfaction level, will need constant updating.

Collecting all these data will help the organization get a better picture of the needs of each customer as well as his or her potential value to the company. If it knows which customers are more valuable, the enterprise will be able to prioritize its competitive efforts, allocating more time and resources to those customers who are likely to yield higher returns. Knowing the needs of an individual customer makes it possible to cater to those needs, and by doing so to lock in the customer's loyalty, increasing his value to the enterprise.[13]

The organization can ask its customers for information during direct transactions. Alternatively, it can buy the information from a third party or get it from an organization targeting the same audience, in exchange for information about other consumers. Consumer information is then consolidated in a computerized database for easy access and updating – particularly after each transaction with the customer.

Capturing, integrating and leveraging all available customer data require a solid process, documented rules and the discipline to use the data effectively. But the rewards are great: consistent, quantifiable, clean data ready to be easily converted into actionable marketing intelligence.

3.3.2 Consumer Data Integration

Ultimately, of course, the purpose of collecting customer information is to develop a closer, more profitable relationship with the individual customer. In many cases this relationship will be facilitated by the availability of information that will make the customer's next transaction simpler, faster or less costly. Remembering a customer's logistical information, for instance, will make the reordering process easier for him. Remembering this type of information will also lead the customer to believe he is important to the company, that his patronage is valued.[14]

However, many arts organizations still operate with a silo mentality, whereby each function (advertising, fundraising, subscriptions, ticketing/box office, client accounting) is relatively autonomous and requires a separate customer information system. Often these systems are not compatible and it is virtually impossible to gather all information about a given customer in a single file. Any extraction and reconciliation requires extensive manual effort.

As a consequence, the organization approaches its customers in a disorderly manner. Patrons receive an uncoordinated array of offers and solicitations – sometimes conflicting – from the marketing, box office and development departments. Also, they have to call a different department for each individual transaction. Furthermore, donors are not acknowledged as such when they book tickets, which probably shakes their loyalty to the organization and weakens their bonds with it.

This needs to change. Every piece of data on an individual customer must be integrated within a single file, to reveal one complete picture of the customer available throughout the organization. Once a customer's identity is fixed, it should be linked to all of his transactions and interactions, at all points of contact and within all of the enterprise's different operating units and divisions.[15] Therefore every database in the organization – front and back office, marketing and finance – must be connected, or, better still, integrated.

A customer who returns to a different part of the organization should be recognized as the same customer, not a different one. In other words, the customer who visits the Website today, makes a purchase in the shop or at the box office tomorrow, and calls the toll-free number next week should be recognized as a single customer, not three separate ones. The data on customer identities maintained in the enterprise's databases must be made available to the people and functions within the enterprise that need access to it. In a service organization, it is especially important that customer-identifying information be made available to frontline personnel.[16]

3.3.3 Application: Royal Albert Hall (London, United Kingdom)

London's Royal Albert Hall, home to the BBC Proms and rock concerts alike, has been a favourite with music lovers for more than 130 years. It stages 365 events and performances per year, for an audience of 1.27 million people.

It recently completed an eight-year, $120-million renovation and redevelopment program. "Towards the end of this program, we started to concentrate on our business practices," remembers Sarah Woods, Director of Customer Relations.[17] "We thought that, after refurbishing this fabulous building, we needed to rethink the services we offered our customers. Central to that vision was the concept of having an integrated system that would record all of our customer details across the organization. We knew this would enable us to capture a single view of each customer's relationship with the Hall."

The Hall held more than one million customer records scattered over several databases. The management team recognized that, if properly integrated within a single information system, this asset could offer much greater benefits than were currently being exploited. Two consecutive dark days enabled the Hall to close its box office and convert data from all the customer records and more than 600 live performances into the new database. "Although the conversion process had been tested a number of times prior to this point, it was still an unsettling prospect for the team," says Woods. "It was an extremely ambitious project and one that could not have been achieved without allocating two full-time staff members to ensure we kept to the deadlines we had set for system set-up, testing and training." At the end of the 48-hour period, all key data had been checked and the box office started selling tickets using the new system.

A new version of online ticket sales went live the following day. "Through integrating our new database with the Web solution, there has been an increase in the amount of customer data collected for marketing purposes," adds woods. "When booking online, customers are given the option of informing the Hall about the type of events they are interested in. This information is recorded as preferences within their account and can be used for e-marketing and direct-mail campaigns."

For many years the Hall had wanted to know its audiences better. It was finally able to do so. A program of audience research was launched, using the data held. "We started by investigating the levels of loyalty within different audience sectors," says Woods. "We ran queries on frequency (how often people attend our events) and on crossover (how many different type of events – ballet, opera, concerts – they attend)."

But the Hall went further. It crossed its internal database with an external database called Mosaic,[18] which consolidates a vast amount of psychographic and behavioural data from 400 different sources (census bureau, house price listing) and aligns them with postal codes. "The data that we hold in our database tells us a lot of information about the ticket purchases at the Hall but doesn't really tell us what our customers do outside," explains Woods. "What magazines do they read? What TV programs do they watch? Where do they shop? What do they buy? Since we know where they live, we can merge Mosaic's data with ours, and answer all these questions. We get a fine portrait of each customer. For example, [a particular] customer comes to the Hall 10 times a year. He likes booking in our restaurants. He attends ballet and opera but also came to see Eric Clapton in 1987. He reads *The Guardian* and shops at Marks & Spencer, probably votes Conservative, etc. It's a very precise tool. Each time we tested it by entering our own personal data, the result was 99% spot on."

Such information proves very useful for preparing direct marketing campaigns, informing future programming choices and attracting potential sponsors. "If we can prove that a large proportion of our audience shops at Marks & Spencer, then we can approach this retail chain to talk about sponsorship," concludes Woods.

3.4 CUSTOMER MODELLING

3.4.1 Principles

"If you can look into the seeds of time, and say which grain will grow and which will not, speak then unto me," wrote William Shakespeare. The challenge facing many companies that amass huge amounts of customer data is to make sense of this information. By modelling these data, a company can do just that, and add the missing link to understanding customers: predicting future behaviour.

Modelling refers to the process of predicting a household's behaviour by comparing the demographic and/or behavioural characteristics of buying households with non-buying households. The modelling process involves searching all the data collected from a company's existing customers (age, income, education, marital status, household configuration, home ownership, etc.) and crossing these variables with consumer behaviour data, in search of correlations. It relies on capturing relationships between explanatory variables and the predicted variables from past occurrences and exploiting them to predict future outcomes.

Simply put, modelling is looking at customer behaviour changes over time as clues to future behaviours. The company analyzes past consumption patterns for each client and follows the evolution of its relationship with him over time, from his first transaction to his defection. This evolution is called "the consumer life cycle." From all the information gathered, the organization can extrapolate a series of standard consumer profiles.

By examining the life cycle of one customer against several consumer profiles, the organization can forecast when, how and how much a customer will buy. For instance, it can predict that a customer who fails to make a second purchase within 30 days of his first is unlikely to return and make any further purchases. Armed with this knowledge, the company can contact the customer before the 30-day deadline with a promotional offer.

Such predictions rarely take the form of absolute statements. They are more likely to be expressed as values that correspond to the odds of a particular event or behaviour taking place in the future.

This process requires the use of powerful software, built on sophisticated statistical models such as regression analysis (see 3.5.2 Forecasting Methods). By analyzing historical information and applying it to current customer data, this software can predict future events, with varying degrees of accuracy, based not just on the amount of data collected but also on the power of the analysis applied to the data. The final form of a model is a statistically derived equation that can "score" each household and rank it according to its likelihood of responding to a particular marketing offer over time.

Targeting only those consumers who are most likely to buy can lead to a substantial increase in response rate, which in turn can lead to a significant reduction in cost per acquisition. Apart from identifying prospects, modelling can help to identify the most effective combination of products, price, promotion and distribution channels for targeting a given consumer.

Modelling is already widely used in identifying the risk or opportunity associated with a specific customer or transaction and making customer decisions. One of its applications is credit scoring, which is used throughout the financial services industry. Scoring models process a customer's credit history, loan application and other data in order to rank-order his likelihood of making credit payments on time.

3.4.2 Customer Lifetime Value

The net value of all these anticipated transactions over the entire history of a customer's relationship with a company is called the customer lifetime value. From a customer's stream of positive contributions, including product and service purchases, an enterprise deducts the expenses associated with that customer, including the cost of maintaining a relationship with him. For instance, relationships usually require some degree of individual communication, via phone, fax, Web, mail, e-mail or face-to-face meetings. These costs, along with any others that apply to a specific individual customer, will reduce the customer lifetime value.

Sometimes the costs associated with a customer outweigh the customer's positive contributions, in which case the customer lifetime value is negative.[19] Unless serving this customer is explicitly part of the company's mission, he should be weeded out.

Customer lifetime value tells a company exactly how much each customer is worth, and therefore exactly how much time, effort and money it should spend to acquire each new customer and retain each existing one. The use of customer lifetime value as a marketing metric tends to place more emphasis on customer service and long-term customer satisfaction than on maximizing short-term sales.

Instead of constantly struggling to acquire more and more new customers, the company can focus on keeping its existing customers longer and selling to them repeatedly. An organization that offers multiple products can concentrate on cross-selling (i.e., selling a given product to the users of another product) or up-selling (i.e., selling a superior version of a given product to the users of an inferior version). For instance, a performing arts organization can identify, among its single-ticket buyers, those who demonstrate potential for greater involvement with the organization and concentrate on turning them into subscribers and then donors.

3.4.3 Application: Stratford Shakespeare Festival (Stratford, Ontario, Canada)

After years of collecting, aggregating and analyzing data on its audience, the Stratford Shakespeare Festival, the largest classical repertory festival theatre in North America, has developed a sophisticated customer modelling system to better allocate its marketing resources and make more informed, wiser decisions.

"We completely re-engineered our business," says Lisa Middleton, the Festival's Audience Development Director.[20] "We analyzed every business process, every rule or practice, and asked ourselves, 'Why do we do this?' We were focused on putting the customer at the centre of our organization and [on not] having him or her be several persons: a member, a single ticket buyer. Our ultimate goal was to have one single view of a customer."

The company has merged customer data, census data and past sales data to create a predictive model of lifetime value of customers. It has segmented its database of 400,000 patrons by using two criteria: value (how much a patron buys) and behaviour (why, what, when, where and how he buys).

The value segmentation variable ranks patrons according to their purchasing pattern. For instance, "preferred" patrons have attended the Festival each year for the past four seasons and are in the top 10% of the spending bracket. The company gives them significant attention, contacting them by mail, phone and e-mail. "Conversion" patrons have attended the Festival once during the past five seasons and have no history before or after that visit. They are usually contacted only via low-cost methods, such as e-mail.

The behavioural segmentation variable allows the company to better understand what offer might get a response from a patron. For instance, "Standard Fair" patrons seek entertainment and attend only the musicals and the popular Shakespeare plays, whereas "Non-Standard" patrons seek artistic stimulation and attend more experimental productions on smaller stages and the lesser known Shakespeare plays. This information allows the Festival to develop "variable printing on demand" (VPOD) mailings by printing customer-specific text and/or images onto the materials sent to a patron according to the segment to which he belongs. It allows it to develop a personalized relationship with each patron by acknowledging what he saw last and to present what he might enjoy seeing next.

By cross-examining and correlating the two variables, the company can model customer behaviour and identify which patrons have the potential to move to a higher spending bracket, and what type of promotional incentive or discount should be sent to them.

3.5 SALES FORECASTING

3.5.1 Principles

"Prediction is very difficult, especially if it's about the future," said the Danish physicist and Nobel laureate Niels Bohr. No marketing strategy is complete without forecasting future sales, yet forecasting entails many paradoxes. The future can be predicted only by extrapolating from the past, but we can be fairly certain that the future will be different from the past.[21] This prediction should factor in uncertainty, which is by definition unpredictable.

It is often said that there are two types of forecast: lucky and wrong. This is why, since ancient times, most oracles have had the wisdom, when predicting the future, to use double-meaning, conditional, obscure and general wording that cannot be checked against reality and to tell people what they want to hear.[22]

But here, too, information technology has greatly improved our ability to forecast and to estimate uncertainty. It is now possible to analyze all past sales and to track, in real time, the actual level of booking for a given event over a lead-in period. By crossing historic sales data with booking data, one can calculate and recalculate future booking demand and continuously update the forecast.

Forecasting has several uses.

To answer "what if" questions. In considering which strategy and tactics to use, the key is to estimate the outcomes of the various strategies and tactics, typically the sales and profit levels. The simplest "what if" question is: What will happen next year if everything remains as it has been in the past? This makes the forecast basically an extrapolation.

To help set budgets. Sales forecasts become the basis of a budget because they specify both sales levels to be attained and, by implication, the resources needed. All pro forma income statements are based on a sales forecast.

To provide a basis for a monitoring system. Deviations from forecasts serve as warnings to management to re-examine a market and their strategy in it. Both positive and negative deviations from forecasts can lead to a better understanding of the marketplace through an examination of the underlying causes.

To aid in production planning. With more companies and their channels moving to just-in-time production and distribution systems with low levels of inventory, accurate forecasting is becoming even more critical.[23]

A good forecasting process starts with a sourcing of the database for historical sales data, in order to analyze sales over time. Any peaks or valleys in the sales series will alert a good manager to try to uncover the factor that has caused such a sharp change.[24] It is also possible to determine seasonal trends (winter vs. summer, midweek vs. weekend, afternoon vs. evening).

Due to the prototypical nature of many artistic products, it is often difficult to build a consistent sale historic. For instance, comparing the sales of one play to those of another might be irrelevant because of the different natures of the two plays. It is then necessary to break down each product into subcategories. For instance, some plays are new while others are from the repertoire; some are avant-garde, others traditional; some are comedies, others tragedies; some have a long run, others a short run; and some feature well-known actors, others up-and-coming actors. It is possible to group these plays into clusters (e.g., new, short-run comedies with famous actors) and compare the sales of the plays in the same cluster through time.

A forecast must also take into account all the factors that can influence sales. These can be grouped into two categories:

1. External factors over which the company has little or no control, such as the state of the economy, demographic changes in the population, costs of basic resources, weather variations and competitive forces.
2. Internal factors, which the company can control. Essentially, these are the four elements of the marketing mix: product, price, place (distribution) and promotion.

Forecasting can be thought of as the process of assessing possible outcomes under likely combinations (often called scenarios) of all these factors.[25] Producing a forecast for each possible combination of factors is a tedious task at best. This process is greatly facilitated by computer analytics.

3.5.2 Forecasting Methods

A large number of methods have been developed for forecasting. Some rely on judgement (of the sales force, of a panel of experts). Some rely on customer feedback using market surveys. Some rely on simple sales extrapolations using past sales data and adding a growth coefficient through time.

Some of the more sophisticated forecasting methods rely on modelling to determine causal relationships between all the variables that can affect sales and sales results.

Regression analysis is the most commonly used modelling method. It consists of analyzing past sales and finding a statistical correlation between variation in sales and each independent factor – whether external or internal. Each factor is then assigned a coefficient, called the lift, which measures its influence over sales variations relative to all the other factors. A big coefficient indicates that this variable has more impact on sales than others. For instance, when processing past sales data for a cluster of plays, one could discover a strong or weak correlation between variations in promotional activity and variation in attendance. The same process can be reiterated for each individual factor. One can go a level deeper, by dividing that promotion factor into subcategories (coupons, two-for-one tickets, sweepstakes) and comparing each with variations in sales.

Today, most personal computers have enough memory and computing power to run a sophisticated, multi-level regression analysis.

Once the lift of each factor is determined, it is possible to create scenarios by combining assumptions for external factors (e.g., the economy will remain stable and the competition's promotional activities will increase by 10% over the forecast period) with decisions concerning internal factors (price increase or decrease, more or less promotional activity), and to forecast sales under these conditions.

One can also assess the accuracy of such a forecast. This is often a function of the quantity and quality of the data collected. Obviously, more accuracy in a forecast is better than less, but at some point the cost of improving the forecast will exceed the benefit.[26]

No forecast is ever completely accurate, but as the French mathematician Henri Poincaré (the father of chaos theory, which has had a great influence on forecasting) said, "It is far better to foresee even without certainty than not to foresee at all."

3.5.3 Application: Steppenwolf Theatre (Chicago, Illinois, Unites States)

Chicago's Steppenwolf Theatre Company[27] has developed sophisticated analysis tools on both the development and marketing sides that can provide immediate feedback on which promotional offer is working and which audience segments are responding to it.

This capability comes in handy at a time when Steppenwolf, like many American theatres, is experiencing an erosion of its subscription base and has to rely more and more on single-ticket purchases – which are much less predictable. "For the performing arts, the big issue on the table is our declining subscription base," says Erin West, Steppenwolf's Marketing Director.[28] "We need to change our relationship with our audience, figure out how to engage them in a new way. We can use technology to do it.

"In the last two years we've spent a lot of time working on predictive models for revenue. In addition to the basic demographic information that we collect at each transaction – addresses, education, gender, income, ethnicity – we're now collecting more subjective information about what they think of the play: overall experience, play content, playwriting, acting, design. We ask them to grade each item on a five-point grid from Excellent to Poor. We're tracking all of those different indicators very closely. We run them against sales to see if and how they correlate with box office revenues. For the overall experience, we're particularly looking at Excellent, because we think it's a motivator for strong word of mouth. This is very preliminary. We're still trying to gather the data and trying to massage it, look at it and see what it tells us.

"What if we could know that when 45% or more of our audience tell us that they had an excellent overall experience for a given play, that means we're going to hit our revenue goal, or more? And if we have 25%, we're going to bring in X number of dollars? That information could give us, in the first few weeks of a run, or even in the first week, an indication that we're not going to meet our revenue goal for the play. We then can change our marketing strategy accordingly. We can decide to brace ourselves and commit additional resources to promoting the play. Or we can decide to cut our losses, stop investing time, energy and money for that play and concentrate on maximizing our chances for the next one.

"We're also integrating other data that factor into revenue over time: which actors are in the play, time of the year, length of the play, known or unknown author and title. We try to find out if there are other indicators behind sales history that would be useful in predicting sales. We also track week-by-week sales for the other plays of the season and historic sales from the five previous seasons. So at a certain point we can say, 'Alright, if this continues at this point, all the other plays are at 30% of their final goal,' and we project it out.

"It adds a sense of predictability. It's not completely refined and it's not an exact science, but we're getting pretty good at it. Basically, at opening we can predict the sales for the nine-week run. For the last play, we were within 3% of where we actually ended up."

3.6 DYNAMIC PRICING

3.6.1 Principles

"The cynic," Oscar Wilde wrote, "knows the price of everything and the value of nothing." Matching price and value has always been difficult, particularly in the arts. The aesthetic value of a cultural product is eminently personal and subjective. Demand/price elasticity for the same cultural product can vary from one consumer segment to another. Demand for a product can also vary from one time period (a day, a week, a month) to another. Lastly, price itself influences value perception: Since all that is rare is expensive, we often assume that what is expensive is rare.

Setting a fixed, standard price for all customers does not allow organizations to take advantage of these variations. If a cultural organization wants to maximize profits, it has to develop a *dynamic pricing* policy. Dynamic pricing sets different prices for the same product according to the consumer segment, consumer behaviour or time of consumption – not according to the costs of producing the product. The organization can set in advance the different prices at which it will sell the same product to different customers. This process is called *segmented pricing*. Alternatively, the organization can negotiate the price with each customer during the transaction. This process is called *negotiated pricing*. Because of the evolution of information technology, each consumer can instantly compare the quoted price with the price of comparable products and make a better-informed decision.

3.6.2 Segmented Pricing

If an organization observes demand/price elasticity variations from one consumer segment to another, it should try to understand the consumer preferences behind these variations and structure its prices accordingly. For example, a performing arts organization can scale its prices for a given show using the following parameters:

Location of the seat relative to the stage. Orchestra, balcony or loge seats can be sold at vastly different prices. This difference is not necessarily correlated to variations in acoustic or visual quality from one seat to another.

Time and date of the show. Demand for a given show can vary according to three factors: time of day (matinee or evening), day of the week (midweek or weekend) and time of year (different seasons, school holidays). A price scale could reflect these variations.

The organization could then divide its inventory into seat allotments for each price category, allowing each customer to purchase the ticket that best fits his budgetary or time constraints. The organization will thus be able to improve attendance and income.

After analyzing past sales, the organization could decide, before the start of the season, to change the seat allotment or the price for each category, to better match demand. For example, if the orchestra seats in the first 10 rows sell better than those in the next 10 rows, the company could raise the price of the first 10 rows or move rows from one category to another.

The company could also entice its customers to buy early by basing its pricing on the date of purchase. These anticipated sales will improve its cash flow.

This approach to pricing is far removed from the traditional "one price fits all" policy.

3.6.3 Yield Management

Until recently, it was difficult – even seemingly impossible – to change the number of seats or the price in each category *during* a season. This is now feasible.

The evolution of information technology and online distribution, the accumulation of consumer information in large databases described above and the manipulation of data with powerful mathematical tools have given rise to a new, scientific pricing technique, Yield Management (also called Revenue Management or Real-Time Pricing).

According to Optims,[29] a French software company, this technique "calculates the best pricing policy for optimizing profits generated by the sale of a product or service, based on real-time modelling and forecasting of demand behaviour per market micro-segment. It offers an excellent solution to the problem of comparing supply and demand thanks to differentiated pricing and systematic control of the inventory for sale in each price category. All players benefit from using this concept: the producer gains in increased turnover and revenue; the end-user enjoys lower prices for the same quality of service."

This pricing method was developed for the airline industry and has spread to other industries, such as tourism and hotels, that share the following characteristics:

- Production capacity is fixed but demand for services fluctuates.
- Variable costs are low.

- Inventory is perishable: Services cannot be stored if not sold.
- Services can be sold through reservation before the date of production and consumption.
- Demand/price elasticity varies from one consumer segment to another, calling for segmented pricing.

Many cultural organizations fit this model, particularly in the performing arts field, and can use this pricing technique.

Implementing Yield Management requires careful planning, to ensure that consumers do not respond negatively to variations in price.

3.6.4 Negotiated Pricing

Theoretically, the maximum profit is reached when each product is sold at the highest price possible, dictated by supply and demand. This point cannot be reached when the price structure is set in advance. To achieve this goal systematically, the enterprise must negotiate the price of each product with each customer, in order to reach each customer's reserve price (i.e., the maximum price he is willing to pay for that particular service).

Negotiated pricing for cultural products goes back to ancient times. Christie's and Sotheby's, the world's two leading auction houses, have been using this method for more than two centuries. In the performing arts, "scalpers" have mastered the technique. Until recently, however, negotiated pricing could not be used profitably for most cultural products. It was limited to rare and expensive items and relegated to the resale or grey markets.

But the Internet has opened up a new era: Online auction sales are growing exponentially. Any cultural organization can now use an auction site such as eBay, the world leader in its field, to sell its inventory. It can also plug auction software into its own Internet site. This is already common practice in the field of sports and could be extended to culture and the arts.

3.6.5 Application: Chicago Symphony Orchestra (Chicago, Illinois, United States)

Historically, performing arts organizations have not been well equipped to react to poor attendance, to take advantage of box office successes and match price with customer value. They usually set their pricing scale at the beginning of one season and review it at the beginning of the next. But setting a fixed, standard price for all customers does not allow them to take advantage of demand variations. Demand/price elasticity for the same cultural product can vary from one consumer segment to another. Demand for a product can also vary from one time period (a day, a week, a month) to another. Lastly, price itself influences value perception: Since all that is rare is expensive, we often assume that what is expensive is rare.

To remedy this, the Chicago Symphony Orchestra, one of the world's leading orchestras, has developed a dynamic pricing policy, setting different prices for the same product according to consumer segment, consumer behaviour or time of consumption – not according to the costs of producing the product.

The CSO uses the wealth of consumer data it collects to understand, anticipate and react to customer behaviour. It then structures its prices accordingly. After analyzing past sales, it can decide, before the start of the season, to change the seat allotment or the price for each seat category, to better match demand. For example, if the orchestra seats in the first 10 rows sell better than those in the next 10, it can raise the price for the first 10 or move rows from one category to another.

But the CSO goes further: It also changes the price for each category *during* the season. It calculates the best pricing structure for optimizing revenues, based on real-time modelling and forecasting of demand behaviour per market micro-segment. This pricing method, Yield Management, was developed for the airline industry.

"We scrutinize the sales for all of our concerts regularly and highlight high-demand concerts as increased revenue opportunities," explains Melanie Kalnins, Manager of Marketing and Sales Analysis.[30] "We look at the number of tickets sold for each price category and the pace at which they sold and compare this to historic trends. We then increase the price of the categories that sell the best and the fastest. We tend to be more cautious at the beginning of a season because we need affirmation that preliminary strong trends will continue. But as the season progresses, we gain assurance that anticipated 'hot' concerts continue to sell, and we decide to raise the price by up to 30%. During our 2005/06 season we increased the prices of 74 concerts (out of over 200) and improved our revenues significantly."

3.7 RELATIONSHIP MARKETING

3.7.1 Principles

This deep knowledge of each customer, the capacity to understand his taste and preferences and to predict future transactions, profoundly transforms the marketing function. In fact, it announces the end of traditional mass marketing – the primary objective of which is to sell products to unidentified consumers, using the same rationale and message for every consumer – and opens a new era of relationship marketing, the primary objective of which is to deepen the relationship between the organization and each customer in order to increase sales.

This customized relationship, from One to One,[31] is in itself a source of wealth for an organization. It builds customer loyalty by tailoring services to consumer profiles, by offering the services when customers want them and by allowing customers to critique the offer made.

The Web permits consumers to submit their specifications online directly to the company. In fact the value a customer adds by inputting his preferences makes the relationship richer and the customized product or service worth more.

Seamless interaction with customers is in no small part a function of customer data integration. "Today, each operator in our call centre can instantly have any individual customer records at his fingertips," explains Sarah Wood of London's Royal Albert Hall. "Our box office now operates what we call a 'one-stop shop.' With one phone call, customers can book their entire experience at the Hall, including ticketing, parking, catering requirements and tours of the building. Box office operators get a single-screen view of the customer, which details their entire relationship with the business, and use this information to offer a more tailored, better-informed service."

But this type of Customer Relationship Management (CRM) necessitates significant investments in software and hardware, an ongoing information-gathering process and the ability to customize offers. To be profitable, it also requires that priority be given to high-potential clients. All of this is not always possible – or desirable – for cultural organizations.

Moreover, a large portion – perhaps a majority – of CRM technology implementations have failed miserably. Some of the problems with CRM implementation are technological, but the vast majority of failures stem from other factors, including a lack of supporting vision, the absence of an overarching customer strategy, failure to align new technological capabilities with management practices and business processes, inadequate education and training of personnel, and insufficient attention to the kinds of success metrics that are more appropriate for a customer-focused strategy.[32]

3.7.2 Total Experience

In a seminal book titled *The Experience Economy,* Joseph Pine and James Gilmore urge every company to find inspiration in the performing arts to improve their customer services. "Work is theatre," they write. "Whenever employees work in front of customers, an act of theatre occurs. Every action contributes to the total experience being staged. Business performances must rival those featured on Broadway. With theatre as the model, even mundane tasks engage customers in a memorable way."[33]

Ironically, by focusing all energies on the art on the stage or on the walls and neglecting all other aspects of customer service, many arts organizations make theatre seem like work. Although a common reason for subscription non-renewal is dissatisfaction with the quality of the art, there are many cases of "audience abuse": arrogance towards customers, inflexible ticket-exchange policies, unannounced seasons, last-minute schedule changes, ticket-issuing mistakes.

While audience reaction to a new play or exhibition is not controllable, the above aspects are: Management can train staff to be customer-friendly, ensure that queries are answered promptly, and so forth.

Some innovative arts organizations have begun to "script" the customer experience beyond the mere attendance at the play. For example, the Pasadena Playhouse in California has recruited a Customer Experience Manager whose mission is to ensure that every aspect of the subscriber experience is complete, fair, uniform, fiscally responsible and artistically interesting. The theatre has adopted a flexible and innovative ticket-exchange policy: Should a patron not enjoy a performance, he or she can go to the box office at intermission and request a credit in the form of tickets to another show during the same season.

"We have to make our guests happy the moment they get in," says David Hawkanson of the Steppenwolf Theatre Company in Chicago. "For instance, we greet every new subscriber with a welcome card placed on their seat for their first visit. We have to make them think, 'This is different – I walked in here and somebody greeted me.' We're constantly telling our box office people, our front of house people – even our parking people: 'You are the first impression. Whatever you do, however you look, however you treat that person...you have more power to create a positive feeling for that evening than the actors on stage do. And if you screw it up on the phone or at that box office window, if you don't have the right information and create the wrong impression, then you screw it up for the actors on stage. They don't have a chance if you didn't do your job well.'

"We want our guests to think of their experience with us as more than a retail exchange. We want to make sure that, when they meet a friend who asks them why they go to Steppenwolf, they can talk about it with a sense of ownership – even if they didn't like our last play. We work constantly at deepening our relationship with them, by improving the whole arc of their experience with us. Increasingly, the arc of experience will start online – when a client buys a ticket, and finish online – when he posts his comments about the play he just saw on our blog. Our Website will be the first and last point of contact."

3.7.3 Application: Metropolitan Opera (New York, New York, United States)

The Metropolitan Opera is a giant in the performing arts field, with 240 performances of 30 different operas every year, 800,000 tickets sold annually, a 4,000-seat auditorium and annual operating expenses of more than $220 million (roughly what it takes to run the next five largest opera companies in the United States).[34]

Until the mid-1990s, despite the Opera's artistic achievements, technical prowess and audience acclaim – or because of them – little attention was paid to marketing, box office management or customer relations. "When you regularly sell 92% or 93% of your seats with very little marketing effort, you can become a little bit complacent," admits Stewart Pearce, Assistant Manager for Operations.[35] "We were not as customer-friendly as many other organizations, and we got away with it because tickets were selling well."

"The box office was not even automated," recalls Smeeta Sharon, former Assistant Manager for Business Affairs.[36] "After a customer requested tickets, he didn't have confirmation that he actually had seats. That confirmation happened when he received the tickets in the mail, many days or weeks later, sometimes after the performance! We didn't even record our financial transactions in a simple, efficient way."

As for customer satisfaction, the company was operating blind. "We had no mechanism to capture the voice of customers and collect their complaints," adds Sharon. "We had no idea how they felt about us. If you don't measure it, you don't know it and you can't change it."

Something had to be done. "We had to make a long-term commitment to improve our customer relations, even if, at that time, our budget was balanced and everything seemed fine," says Pearce. "With 93% tickets sold, we didn't have a lot of unused capacity to face our growing expenses. We couldn't keep on raising our ticket prices by 6% or 7% a year as we had done in previous years. So we had to market more smartly, more cost-efficiently, and leverage our ticket sales into increasing fundraising."

But how to convert frustrated patrons into donors? "If we had been rude to a client, why would he give us another dime?" asks Sharon. "He might still buy tickets from us, because he loves opera, but he'll certainly not contribute to our fundraising campaigns. We had to delight our customers so that they would remember their great experience with us when we asked them for support. So I started to read every letter sent by customers...and realized that their experience was often less than delightful."

But where to start? "We reviewed every operation of our box office, marketing and development functions," recalls Sharon. "We tried to remove bottlenecks and duplication of effort." A separate, customer-centric department called Customer Care was created to act as a single interface between the customer and the company. It would regroup full-service agents capable of handling every transaction for a customer. Inversely, all the Met's departments had to go through Customer Care to contact the customer.

"If the marketing department decides to over-market to these constituents and it clashes with our fundraising efforts, then it's the role of the Customer Care Department to say, 'No, you're annoying the constituent, you're sending too many mailings, you're making too many phone calls,'" explains Sharon.

To be fully efficient, Customer Care employees needed to have instant access to all information about a patron. But the data available was highly fragmented and resided in different systems. "We had three separate databases," says Sharon, "a ticketing one, a subscriber one and a donor one, plus a multitude of Excel spreadsheets that people kept on their personal computers. Any extraction and reconciliation required extensive manual effort. Bridging these databases would always be a problem. We needed to operate all our functions from a single platform, with a single database."

The Met soon realized that no existing system integrated the ticketing, fundraising and marketing functions on a single platform, and none could handle its 1.5 million customer records and its transaction volume. The Opera would have to develop a brand new system in-house.

This revolutionary new software is called Tessitura. Because it uses a single database of information, Tessitura empowers the user organization to record, track and manage all contacts with its constituents, conduct highly targeted and cost-efficient marketing and fundraising appeals, handle all ticketing and membership transactions, and provide detailed and flexible performance reports. Marketing and fundraising targeting can be based on any criteria, attribute or data point in the system. All data are maintained in complete synchronization.

"Tessitura tells us something about almost every Met patron," says former General Manager Joseph Volpe, "from pattern of residency to pattern of ticket-buying, history of charitable contributions to, as far as we can determine, spending habits. There are 1.5 million names in the database, of which somewhere between 300,000 and 400,000 are active. It is an indispensable tool for finding out who our vast audience really is. The moment the Met's telephone order-taker hears your name or phone number, we know who you are."[37]

A key benefit of the system is the ability of relationship managers to have a real-time full view of each constituent – history, preferences and all contacts with the institution. "Tessitura allows you to capture customer issues," explains Sharon. "If you call and complain that the podium light in the pit was too high and disturbed your viewing of the opera, it will be recorded as a customer issue. And at the end of the season, when we run a report to see all customer issues under auditorium experience, the report will show how many people complained about the podium light or bad ventilation or bad seats. We can ask the house manager to fix these problems. If you sat next to someone you didn't like at a fundraising dinner and you happened to mention it to someone on staff, their job is to add it to your record the next day so that we don't seat you next to this person again. Tessitura allows you to capture every single contact point with a client."

Because of its real-time Web-ticketing application, the system drastically reduces data entry – which is done by patrons themselves when they buy tickets online – and reduces the time and cost of the transaction. It helps to reduce the costs of direct mail by better targeting marketing materials, and it enhances client satisfaction by responding immediately to all customer service issues.

Tessitura has been adopted by many arts organizations, including the Sydney Opera House, Royal Albert Hall, the Stratford Shakespeare Festival, Steppenwolf Theatre and the Chicago Symphony Orchestra.

3.8 BUILDING TRUST

Gathering more and more detailed information about current and potential clients raises some concerns about privacy, particularly when an organization seeks to collect financial, medical or personal information.

Gathering this information through the Internet is even more problematic, since the public, open nature of the Internet does not guarantee confidentiality. Internet communication has an inherent capability to mine, store, organize and transport information about those who use it – often without their knowledge. "You have zero privacy on the Internet anyway. Get over it," warns Scott McNealy, CEO of SUN Microsystems.[38]

Moreover, Internet users can fall prey to criminal activities, such as unauthorized disclosure (interception of sensitive transaction information when transmitted without prior encrypting), data alteration (the altering of customer data during transmission to a company's database) and "phishing" (posing as a legitimate business via digital communications for the purpose of extracting information such as social security numbers, credit card numbers or bank account numbers).

In many countries, these concerns have led to legislation regulating the gathering process in order to protect consumer privacy – particularly that of minors.

To fully exploit the potential of applying new information technology to the marketing function, an organization must not only conform to all regulations in place but also define and publish a clear code of ethics, to reassure its clients. Each consumer should have the right to know what type of information has been gathered about him, to know what the organization plans to do with the information, to refuse to have the information shared with any other organization, to have full access to the information and to be able to modify the information. If an organization breaches its code of ethics, it will lose the trust of its customers.

Lose customer trust and everything is lost.[39]

SUMMARY

The information technology revolution (particularly the combination of the Internet, database systems and computer analytics) has transformed the marketing function for the arts, bringing a level of sophistication previously reserved for bigger – and richer – industries. It has left few if any aspects of the marketing mix untouched, transforming the very notion of artistic product and its pricing, promotion and distribution strategies. For companies it brings the promise of more efficiency and for customers the promise of better service.

This chapter explored and illustrated some of the most salient innovations brought by the integration of information technology in the marketing process:
E-marketing (i.e., use of the Internet for direct marketing purposes) allows for more accurate targeting than traditional advertising; it is also more cost-efficient than mailing or telemarketing.
Online distribution is often faster and less costly than the traditional "bricks and mortar" alternative, and it allows for better service, tighter control and finer customer data harvesting.
Customer identification becomes instantaneous with information technology. All the information collected during a direct online transaction can be accessed as soon as the customer contacts the organization, providing a clear picture of what the customer needs or what value he might represent.
Customer modelling is looking at customer behaviour changes over time as clues to future behaviours. The organization uses all the data it has collected to analyze the consumption patterns of each client and follow the evolution of its relationship with him, from his first transaction to his defection.
Sales forecasting is greatly improved by the information technology revolution. It is now possible to analyze all past sales and to track, in real time, the level of bookings for a given event over a lead-in period. By crossing historic sales data with booking data, the company can calculate and recalculate future booking demand and continuously update the forecast.
Dynamic pricing sets different prices for the same product according to each consumer segment, consumer behaviour or time of consumption, instead of according to the cost of producing the product.

This deep knowledge of each customer – the ability to truly know his tastes and preferences and to predict his future transactions – has transformed the marketing function. In fact, it announces the end of traditional mass marketing, whose primary objective is to sell products to unidentified consumers, using the same rationale and message for every consumer, and the beginning of the era of relationship marketing, whose primary objective is to intensify the relationship between the organization and each customer in order to increase sales.

Notes

1. S. Rapp, *The Great Marketing Turnaround* (Upper Saddle River, NJ: Prentice Hall, 1990), p. 50.
2. P. Kotler and G. Armstrong, *Marketing: An Introduction* (New York: Prentice Hall, 2005), p. 459.
3. For more information, see P. Ravanas, "Born to Be Wise: The Steppenwolf Theatre Company Mixes Freedom with Management Savvy" (*International Journal of Arts Management*, Vol. 8, No. 3, 2006), p. 64–73.
4. A. Singer Kaissar, *Theatre Subscriptions in a Changing World* (www.artsmarketing.com), May 2005, p. 2.
5. Interview with the author, December 9, 2005.
6. Interview with the author, February 15, 2007.
7. A blog is a user-generated Website where entries are made in journal style and displayed in reverse chronological order.
8. Interview with the author, August 7, 2006.
9. M. Madlberger and A. Sester, *The Last Mile in an Electronic Commerce Business* (Vienna University of Economics and Business Administration, 2005), www.wu-wien.ac.at.
10. E. Carr, *Websites for Culture* (New York: Patron Publishing, 2005), p. 13.
11. S. Krug, *Don't Make Me Think* (Berkeley, CA: New Riders Publishing, 2000), p. 11.
12. Interview with the author, August 7, 2006.
13. D. Peppers and M. Rogers, *Managing Customer Relationships* (Hoboken, NJ: John Wiley, 2004), p. 114.
14. Ibid., p. 110.
15. Ibid., p. 113.
16. Ibid., p. 94.
17. Interview with the author, August 7, 2006.
18. For more information about Mosaic: http://www.experianmarketingsolutions.co.uk.
19. Peppers and Rogers, *Managing Customer Relationships,* op. cit., p. 117.
20. Interview with the author, August 8, 2006.
21. S. Makridakis, *Forecasting, Planning and Strategy for the 21st Century* (New York: Free Press, 1990), p. 66.
22. Ibid., p. 66.
23. D. Lehmann and R. Winer, *Analysis for Marketing Planning,* 6th ed. (New York: McGraw-Hill, 2005), p. 179.
24. Ibid., p. 194.
25. Ibid., p. 180.
26. Ibid., p. 181.
27. For more information, see Ravanas, "Born to Be Wise," op. cit.
28. Interview with the author, February 15, 2007.
29. For more information on Optims: www.optims.com.
30. Interview with the author, August 8, 2006.
31. D. Peppers and M. Rogers, *The One to One Future* (New York: Currency, 1993), p. 5.
32. Ibid., p. 192.
33. J. Pine and J. Gilmore, *The Experience Economy* (Boston: HBS Press, 1999), p. 104–108.
34. For more information, see P. Ravanas, "A Quiet Revolution: The Metropolitan Opera Reinvents Client Relations Management" (*International Journal of Arts Management,* Vol. 9, No. 3, 2007), p. 78–87.
35. Interview with the author, September 28, 2006.
36. Interview with the author, August 7, 2006.
37. J. Volpe, *The Toughest Show on Earth* (New York: Knopf, 2006), p. 262–263.
38. P. Sprenger, "Sun on Privacy: 'Get Over It'" (*Wired Magazine,* January 26, 1999).
39. Peppers and Rogers, *Managing Customer Relationships,* op. cit., p. 81.

PLAN

Chapter 4

Planning for Fundraising and Sponsorship

By J. Dennis Rich

OBJECTIVES

- Identify the rationale for including fundraising and sponsorship in marketing planning.
- Evaluate principles for effective fundraising and sponsorship.
- Describe the steps in developing effective fundraising and sponsorship plans.
- Understand the process of evaluating successful fundraising and sponsorship.

INTRODUCTION

Thirty years ago, in an article titled "Culture by the Pound," the anthropologist Davydd J. Greenwood wrote:

> Worldwide, we are seeing the transformation of cultures into "local color," making peoples' cultures extensions of the modern mass media. Culture is being packaged, priced, and sold like building lots, rights-of-way, fast-food, and room service, as the tourist industry promises that the world is his/hers to use. All the "natural resources," including cultural traditions, have their price, and if you have the money in hand, it is your right to see whatever you wish.[1]

Across much of the cultural sector, especially in Europe, policies have been created to prevent such commodification of the arts and culture. As a result, European arts policy mandates public support for culture. In the United States, support for the arts is endorsed, but arts policy places an emphasis on diverse sources of support for cultural activity, with little dependence upon public funding. In both cases, a principal belief is that art and culture are a nation's heritage and therefore deserve to be, and indeed must be, supported through direct subsidy or indirect government assistance. Put another way, the collective heritage of a nation merits collective support. Such support frees artists and cultural institutions to pursue artistic endeavour.

Unfortunately, the economic realities of the 21st century make government funding increasingly difficult. The fact is that national governments are faced with other spending priorities. Education, health and maintaining infrastructure are viewed as more critical. Socio-economic factors have reduced government budgets and, at the same time, broadened the competition for subsidy.

In recent years, the average federal expenditure for culture in Europe has dropped by 1% per annum, while regional and local subsidy has increased by approximately 1.5%. However, this small net increase does not begin to offset an annual inflation rate of 2.6% or higher. In the United States, public sector support for culture and the arts has historically been limited; for example, government support from all sources for the performing arts averages 5.5% or less, while in Canada support for all the performing arts from the different levels of governments comes to as much as 40%.[2]

What is described as the "income gap" in William J. Baumol and William G. Bowen's 1966 study of the performing arts in the United States is becoming a reality for culture and the arts in general. More and more, financial challenge is becoming a worldwide way of life in cultural management.[3]

So, how can cultural institutions continue to pursue their missions and avoid the commodification described by Professor Greenwood? It is clear that rising costs will not be balanced by rising governmental subsidy at any level.

At the same time, a significant reduction in expenses is at best improbable, given the determination not to compromise artistic excellence.

It is also clear that there is no miracle solution to the income gap. There are no vast untapped sources of wealth.

The task of the cultural manager is to make the work of the artist possible. More and more, this means careful planning to secure the financial means required for artistic production.

The diversification of funding sources offers greater opportunities for artists and cultural institutions. Clearly, there are risks that lie along this road, but well-crafted fundraising and sponsorship plans are part of the path to flourishing arts institutions.

Tyler Cowen, in his essay "The Arts in a Market Economy," observes that "governments, even democratic ones, tend to favor the cultural status quo that put them in power, or to shape a new status quo that will cement their power."[4] Consider the proposition that a variety of funding sources offers the arts more freedom than an exclusive reliance on public funds. Such diversity offers greater stability for the arts, because it ensures that in periods of austerity or political transition, government cutbacks will not subject arts agencies to severe financial pressure. Further, diversity of funding sources reduces the concentration of government power over arts policy and gives arts institutions a freedom of action they would not have if they were dependent solely upon the government for their funds.

4.1 UNDERSTANDING THE PRIVATE SECTOR SUPPORT MARKET

Funding by individuals, foundations and corporations, in addition to government subsidy in support of culture, demands that arts managers attract donors' attention and cultivate their goodwill. While such a course of action is not without peril, it will result in a broader constituency and a broader base of support for the arts. Moreover, because of the backing of donors, the artist and the arts institution will remain relatively free from the distractions of the marketplace.

This may sound like a point of view unique to the United States, but in their recent study, *Financing Culture and the Arts in the European Union,* Arjo Klamer, Lyudmilla Petrova, Anna Mignosa and the Stichtung Economie and Cultuur write:

> The arts exist by virtue of the people involved, interested and committed. The challenge is to increase the involvement and enlarge the circles of the worlds of art and culture. One way to do that is to develop alternative ways

> of financing the arts. In particular, an increase in private support – individual contributions, donations, sponsorship – is called for without its reducing government support. People should be aware that the arts are important and that they have good reasons to support them. To persuade people and organizations to contribute more to the arts, cultural organisations will have to adjust their structure, focusing more on community building and marketing. Governments could stimulate these adjustments by matching private support and by legislating fiscal incentives. The purpose of these policies is not to interfere with public support but to swell its value in the sector of the arts and culture.[5]

The authors also observe that the way in which a cultural good is financed will affect its valuation and functioning. Is the income earned, government-subsidized, sponsored or donated? The manner of financing clearly has bearing on the organization of cultural institutions. There are three main sources of funds for the arts. Each has its own rationale, and each requires different approaches to public sector support:

Financing by means of the government. To qualify for government support, cultural institutions and artists have to demonstrate that their activities meet qualitative and quantitative standards stipulated by politicians and government agencies. They also have to account for their activities. If government is the major source of support, the cultural institution is more or less compelled to operate in the government circuit.[6]

Financing by means of the market. Alternatively, cultural institutions and artists can sell their work on the market. This includes the sale of their works to companies in so-called sponsored deals. In market-type arrangements, quid pro quo prevails. Accordingly, cultural institutions have to supply something that the other party is willing to pay for. Needless to say, this type of financing requires an effort quite different from that which is necessary to apply for government funds.[7]

Financing by means of the third sphere. Cultural institutions can also generate funds by means of donations from individuals or private institutions. These donations come in the form of money, work or time. In European history, the Maecenas provided a form of financing typical of the third sphere. Nowadays, foundations assume that role. Financing by means of the third sphere prevails in the Anglo–Saxon tradition and appears to be on the rise in Europe. For this method of financing, yet another approach is called for, one focusing on the idealistic and special character of the arts and on its vulnerable character.[8]

It is unusual to find arts activity, outside of purely commercial activity, that is funded in only one way. So, one of the challenges facing arts managers is functioning effectively in multiple funding spheres – hence the need for *planned fundraising* and *planned sponsorship* activity.

Again, it is important to keep in mind that different funding sources have different points of view and different expectations. Contributions, whether from individuals, foundations or corporations, are philanthropic in nature, while sponsorship is a promotional initiative offered in exchange for publicity, advertising or other benefits. The greatest challenge, when pursuing private-sector funding, is securing financial support without endangering artistic integrity or the institutional mission. It should surprise no one that the business community pursues profit or that foundations provide subsidy to institutions and activities that are congruent with their own mission and goals. The danger is what Stefan Toepler calls "mission or goal displacement."[9] The challenge is to pursue private-sector funding while avoiding the temptation to move away from educational or artistic offerings and towards entertainment-oriented offerings so as to remain financially stable.

4.1.1 The Case Statement[10]

With both fundraising and sponsorship activities, the organization begins by outlining and defining its case for support. The case statement presents evidence of all the reasons why a corporation, foundation or individual should support the organization and its activities.

A strong case statement is a detailed, passionate, intelligent and convincing narrative that explains to potential private-sector supporters why the organization is worthy of their backing. Further, the case statement serves as an internal document, ensuring that everyone associated with the institution understands the mission and work of the organization and can communicate it uniformly to others. All or parts of the case statement should be used to create various documents, funding requests and sponsorship proposals. The information will be adjusted and fine-tuned as needed to fit the purpose of each communication. Case statements are a part of planned fundraising and sponsorship activity. An effective case statement includes:

- the organization's history
- what the institution or organization is – its purpose and mission (why it is important)
- what it plans to do or become (whom it serves, and how)
- why it deserves support – how it benefits society

The case statement may be used as part of a formal proposal, as the basis for a brochure or as background material for a conversation with a prospective donor or sponsor.

4.1.2 Donors

As noted in *Marketing Culture and the Arts,* contributed income may come from any of three private-sector sources:

- individuals
- foundations
- corporations – businesses

The importance of private-sector funding is growing. In 2006, Arts and Business, an organization that supports business and arts collaboration in the United Kingdom, found that 83% of cultural organizations that were polled could not have attracted the audiences they did without private backing.[11]

Individual donors provide assistance based on personal taste and commitment. Only individual donors support culture and the arts out of pure commitment to the organization they choose to support. Long the main source of private support in North America, funding from individual donors is now becoming an important source in Europe. In the United Kingdom, for example, 54% of private-sector funding for culture and the arts comes from individuals.[12] In North America, especially in the United States, the percentage of individual support is even greater.[13] Donors usually are rewarded with some form of recognition, but this does not generally motivate a philanthropic gesture.

Foundations exist to give away money, but they decide how much to give, to whom and when based on carefully prepared policies and procedures connected to the foundation's mission. They seek a match with their own objectives from among the applications received.

Corporations, on the other hand, have a history of following a *quid pro quo* exchange. They describe this as "enlightened self-interest." In fact, it may just be self-interest, because corporate donations require justification in terms of a visible benefit to the company. When the company has stockholders, they too must be convinced of the legitimacy of such giving. This may partly explain the rise in corporate sponsorship activity and the decline in corporate donations.

All donation decisions are made by people, whether it is the minister of culture, the executive director of a foundation, the head of a corporate giving committee or a private individual. And people need to feel a connection with the cultural institution they have chosen to support. Philanthropy is first and foremost a human activity, and "friend raising" precedes fundraising. The connection-building process requires a system of *cultivation* of prospective donors, *solicitation* of donations and *recognition* or acknowledgement of gifts. The process should be ongoing: cultivation – solicitation – recognition – cultivation.

4.2 THE FUNDRAISING PLAN

Of course, strategies for fundraising must be developed by cultural institutions based on their particular situation and need. While it is not possible to prescribe an exact method for planning fundraising activity, an effective plan will have the following components:

- Case Statement
- Situation Analysis – strengths, weaknesses, opportunities and threats, competition for funds
- Fundraising goals and revenue targets, by source – individuals, foundations, corporations, new and renewing donors
- Identification of prospective donors – prioritized by likelihood of success
- Fundraising tactics or methods
- Budget, showing both income and expenditures
- Timetable/Calendar, with deadlines for activities and for achievement of goals
- Staffing requirements
- Information management strategies
- Evaluation plan

4.3 SPONSORS AND SPONSORSHIP

Sponsorship is the term used to describe a relationship between a sponsor and an event, agency or property in which the sponsor pays a cash or in-kind fee in return for access to the exploitable commercial potential associated with the event, agency or property.[14] It is part of a strategically planned promotional effort.

A related term is *Cause Related Marketing*. This is a strategically planned *promotional* effort designed to increase a company's sales or improve its position in the marketplace through actions that also benefit a not-for-profit organization. Generally, this means that when consumers purchase the company's product or service, the company donates to the not-for-profit organization.[15]

Cause related marketing and sponsorship are not simple corporate contributions. Corporate contributions are donations not associated with any marketing program (see Table 4.1). They are generally handled by different people from those who make decisions about sponsorship and cause related marketing. Cause related marketing and sponsorship involve the marketing, communications, promotion or public relations functions of a company. Consider the definitions again. Both sponsorship and cause related marketing are strategically planned *marketing* efforts designed to increase a company's sales or improve its position in the marketplace through actions that also benefit a not-for-profit organization. *Strategically planned marketing effort* is the key idea here. It means that sponsorship and cause related marketing are based on a strategy and a plan – not just coincidental activities. No such effort should take place without a carefully constructed plan. The two partners

Table 4.1
Sponsorship and Charitable Contributions Compared

	Sponsorship	**Charitable Contribution**
Publicity	Highly public	Usually with little fanfare
Source of funds	Typically from marketing, advertising or communications budget	From philanthropic or charitable contributions budget
Accounting	Written off as a full business expense, similar to a promotional or media placement expense	Write-off is limited by tax laws regulating charitable contributions; as a result, accounting/tax considerations are unlikely to influence the way a corporation designates funding of a not-for-profit organization
Objectives	To sell more products or services; to increase awareness within markets and among distant stakeholders (customers, potential customers, geographic community)	To be a good corporate citizen, to enhance the corporate image with closest stakeholders (employees, shareholders, suppliers)
Partner/ Recipient	Events, teams or cultural organizations, projects, programs; a cause is sometimes associated with the undertaking	Larger donations are typically cause related (education, health, disease, disasters, environmental) but can be cultural, artistic or cause related; sometimes funding is specifically designated for a project or program, while at other times it is provided for operating budgets

Source: The Sponsorship Report. http://www.sponsorship.ca/p-issues-callit.html.

– the sponsoring company and the not-for-profit agency or event – should draw up the plan together.

In the 1990s a new concept of sponsorship and cause related marketing began to emerge. Short-term sales-related promotional sponsorships began to be replaced by sponsorship and the integration of cause related marketing into the very identity of companies. In this new form of strategic support, not-for-profit events, organizations or causes are associated with a particular brand,[16] as part of a comprehensive, integrated marketing approach.

Companies that sponsor artistic and cultural events are looking for prestige advertising vehicles. Their hope is that the popularity of the group or event, and the public's affection for it, will be transferred to the sponsor. Usually, funds are drawn from the company's advertising or public relations budget. The public targeted by cultural companies is an extremely lucrative source for commercial enterprises seeking a market segment with strong purchasing power. In fact, several companies in the cultural sphere serve a broad-based market and are therefore perceived by the private sector as excellent promotional vehicles. Faced with media fragmentation, companies are looking for new ways to get their message across in a cost-effective and competitively successful manner.[17] Sponsorship can meet this need.

Over the past decade, sponsorship has been the fastest-growing marketing communications tool. Worldwide, sponsorship expenditure is expected to reach $16.8 billion in 2008.[18] In North America, cultural sponsorship has long been an accepted form of marketing communication and cultural support. In Western Europe, it began to take root in the late 1970s. At first, corporations in Germany, Spain, the Netherlands and France limited their support to art collections. They gradually extended and broadened their assistance, subsidizing the visual arts (collections and exhibitions), heritage (restoration of monuments and buildings), and theatre and classical music (sponsorship of touring for well-known companies). Corporate support and sponsorship of the arts are still fairly low in Europe, but they are growing rapidly.

In the collaboration between cultural agencies and companies, arts managers need to know that the company is engaged in strategic philanthropy designed to enhance its image or increase sales.

Some arts and cultural agencies have a problem with this. Their leadership seems to believe that helping a company make money is inconsistent with their not-for-profit mission. However, as long as the activities of the sponsoring company are ethical and legitimate, and as long as the arts agency is doing nothing to violate its tax-exempt status, there is nothing wrong with participating in a partnership that benefits both partners. The definition of sponsorship suggests collaboration and makes it clear that that sponsorship and/or cause related marketing scheme must also benefit the not-for-profit agency. It may be unrestricted funds, financial support for a particular program, in-kind contributions or increased public awareness. Whatever the form of support, it should be something that advances the mission of the arts agency.

Successful corporate sponsorships are a little like successful marriages. Both depend on finding the right partner – one with similar interests and goals. It takes hard work to ensure the long-term success of a partnership.

Today, companies all over the world are seeking such benefits. In Ireland, for example, companies invest in sponsorship as a marketing tool for a variety of reasons (see Table 4.2). Interestingly, the idea that the company must stand for something is not one of them.[19]

Table 4.2
Main Reasons for Undertaking Sponsorship in Ireland[20]

Rank	Current Sponsorship Objectives	% of Sponsorship Managers
1	Improve community relations	59
2	Promote corporate image	57
3	Increase brand awareness	54
4	Media exposure	39
5	Increase brand image	37
6	Improve PR	37
7	Long-term sales	30
8	Entertain customers	26
9	Staff morale	2
10	Other	6

4.4 SPONSORSHIP AND THE CONSUMER

Consumers understand that companies are in business to make a profit, but today they also expect them to be involved. The *Cone-Roper Cause Related Trends Report*[21] reveals that the public view of sponsorship is overwhelmingly positive. It also reveals that sponsorship can influence what and where consumers buy – 80% of consumers now find it acceptable for companies to engage in cause related marketing, up from 66% in 1993.

Leaders in the sponsorship field interpret these data to mean that cause related marketing and sponsorship have become necessities for brands seeking a strong relationship with customers and communities.

The Cone/Roper report also shows that socially and politically active consumers – a key group for marketers – are especially receptive to companies that link with social issues. This group is likely to switch brands or switch retailers to support a company that is associated with a good cause. Consumers also report that they expect companies to support a cause over time and in a substantive way.

4.5 MEASURING THE VALUE OF SPONSORSHIP

For companies, the value of a sponsorship arrangement depends upon the number of visitors or participants, the location of the company's logo on promotional material for the event, visibility of the sponsor at the event, potential media coverage, image, social impact and the commercial potential of the event.[22] Sponsorship, then, can be measured with these criteria in mind. It also can be evaluated by measuring awareness or attitude changes towards the sponsor's product or services, quantifying sales results, and comparing the value of sponsorship-generated media coverage to the cost of equivalent advertising space or time.[23]

A cultural or artistic enterprise should promote the win-win aspect of sponsorship. Requests or applications for sponsorship should take into account not only the organization's need for support but also the potential benefits to the sponsor. To this end, it is important for cultural marketers to:

- think like a marketer and identify the marketable assets of the cultural or artistic enterprise such as name, community, board members, and public awareness of the enterprise and its programs
- learn as much as possible about the corporation by consulting its annual reports, viewing its advertisements and promotional materials, and reading the business press
- attempt to identify the corporation's target market
- address the corporation's needs by demonstrating that sponsorship will help it sell more products or services while at the same time benefiting the arts enterprise
- identify the primary decision-maker for sponsorships and cause related marketing, such as a marketing and/or sales executive.

While investment in sponsorship is on the rise, sponsors say the quality of sponsorship proposals is declining. And, increasingly, companies are measuring the impact of sponsorships using hard numbers. This means that sponsorships and cause related marketing partnerships must be more carefully planned.

- A good proposal *sells benefits, not features.* Cultural managers tend to be very proud of their program, venue or event. As a result, their proposals often describe the opportunity – the merits of the cause, the artistic excellence of the festival, concert or exhibition, the economic impact of the event – rather than the benefits to sponsors. Sponsors do not buy causes, events, exhibitions or performances. They buy promotional platforms that will help them sell products or services.
- The successful proposal addresses the needs of the sponsor, not the needs of the cultural agency. Sponsorship applications frequently emphasize the agency's need for money. However, corporations that are interested in sponsorship are not motivated by a cultural agency's need for money. They want to know *what's in it for them.*
- The successful proposal is tailored to the sponsor's business category. Different sponsors require different benefits. For example, a proposal addressed to an insurance company might focus on access to the arts agency's mailing list or board of directors, while a proposal addressed to a soft-drink distributor might discuss onsite visibility and sales rights.
- The successful proposal includes promotional extensions. There are two types of sponsorship benefit. Automatic benefits come with the deal and require no further action on the part of the sponsor. These include sponsor identification on collateral materials and onsite signage. The second type of sponsorship benefit rests on the sponsor's ability to build upon the sponsored institution, event or performance through trade, retail and sales extensions. Today, automatic benefits rarely provide sufficient return to justify the time and expense of sponsorship. An effective proposal shows how a cultural property or event can be used as the unifying element or theme for media advertising and sales promotion. It is not enough to provide a prospective sponsor with a checklist of automatic benefits. The proposal should include a menu showing the corportion how it can capitalize on the investment.
- The successful proposal minimizes the risk to the prospective sponsor. It is much easier for a marketing or communications executive to authorize a media purchase than to authorize a sponsorship. A proposal can minimize risk by including guaranteed media as part of the package. It can also list reputable co-sponsors. This will tell the prospective sponsor that the opportunity being offered has been favourably reviewed by other corporations.

- The successful proposal includes overall benefits. The cultural agency should present itself in terms of its total impact on achieving the sponsor's objectives instead of isolating one element, such as media coverage. The idea here is that when it comes to benefits to the sponsoring company, the whole is greater than the sum of its parts.
- Finally, the successful proposal offers the corporation the opportunity to form an alliance that will bring it resources that it would otherwise be unable to access.

4.6 NEGOTIATING THE SPONSORSHIP AGREEMENT

When negotiating a sponsorship agreement, the cultural enterprise must keep in mind at all times that a business relationship is being established. For the sponsor, the event or enterprise is primarily a promotional tool. For the cultural organization, the sponsor is like a client who must receive certain benefits in exchange for money. Both parties need to feel they are getting their money's worth. This means that a written contract is imperative. The contract should include[24]:

- the goals of each party
- goods and services offered by each party
- the geographic area covered by the agreement
- the starting and ending dates of the agreement
- creative specifications, such as allowable use of logos, names and images, which partner controls broadcast advertising, and the meaning of the word "official" when used in advertising and promotional materials
- details on how funds will be accounted for and what portion of sales will accrue to the arts enterprise
- the legal/financial system that is in place to track and distribute funds

The cultural enterprise will have to develop a system for examining and analyzing the sponsorship or cause related marketing campaign from start to finish. It must keep track of the campaign's goals and consumers' perceptions and communicate these to the sponsor. The sponsor will need as much information as possible in order to determine whether its objectives have been met. By providing such information, the arts enterprise will enable the sponsor to evaluate whether its investment has been worthwhile.

After the event or at agreed-upon intervals, evaluative meetings should be held with the sponsor to assess the results from the perspective of both parties. Finally, time should be taken to celebrate and to thank the sponsor for its service in support of the arts enterprise and the community.

4.7 PITFALLS AND DANGERS OF SPONSORSHIPS (FROM THE PERSPECTIVE OF THE ARTS AGENCY)

It would be foolish to believe that sponsorship or cause related marketing is without risk. Arts managers should discuss sponsorship with their key stakeholders, including the board of trustees, staff and artistic leadership. Possible negative repercussions must be discussed. In any sponsorship or cause related marketing venture, the arts enterprise should focus first on its mission. While sponsorship and cause related marketing generate income for the arts institution, the corporate partner derives publicity, image enhancement, the support of volunteers from the arts agency in promoting the venture, and increased sales.

With sponsorship and cause related marketing, the most frequently asked question is whether corporations profit unduly. There is much discussion today about sponsorship and possible conflict of interest. Simply stated, are sponsorship and cause related marketing ethical?

From the perspective of the cultural manager, the reputation of the arts enterprise is a fundamental concern. The arts enterprise must determine whether its mission and its nature will be diminished or fundamentally changed if it forms an association with a for-profit company. Under no circumstances should it become associated with a sponsor that is a competitor of a key stakeholder such as a donor. It is equally important to think through "exclusions" – companies whose products or services appear to conflict with the mission of the arts enterprise. For example, it would be advisable for a children's theatre not to accept a sponsorship from a tobacco company.

Cultural enterprises might consider developing a policy on sponsorship. Such a policy should:

- stress the partnership nature of the sponsorship
- define sponsorship as a business arrangement that is distinct from a contribution
- state the arts enterprise's commitment to the concept in positive terms, as sponsors prefer an enthusiastic approach to one that says, "We will take your money but we don't approve of you"
- clearly state exclusions – those corporations that the agency will not do business with
- spell out the rights of both partners
- include an escape clause whereby the arts agency can cancel the arrangement should the sponsor's activities run counter to the

agency's mission[25]

Sponsorship agreements should always be in writing.

As governments throughout North America and much of Europe move to cut support to arts organizations, corporate sponsors are moving in to fill the breach. Some mega-sponsors have emerged. This has resulted in the renaming of some cultural venues.

As competition for support heats up, corporations are expecting greater recognition for their efforts. In fact, more and more companies are moving away from simple philanthropy and towards sponsorship with a marketing edge. Companies are taking greater care to target and select arts institutions that are well matched with their own demographics. In other words, sponsorship is not about having a good heart!

This does not mean that arts enterprises should respond by engraving the name of a sponsor on the stage floor or inserting it into a play. But arts managers need to seek a balance between outright intrusion into the artistic product and mission and tasteful recognition of a sponsor's contribution.

What can be said with certainty is that arts institutions are increasingly being told to look for alternative sources of funding and to step up their private-sector support. It would be foolish to think that sponsorship and private-sector support are without risk. Arts managers need to discuss sponsorship, including its possible negative repercussions, with their key stakeholders. In any sponsorship or private funding venture, the arts enterprise must focus first on it mission.

SUMMARY

Today, the financing of culture and the arts tends to come from multiple sources, each of which calls for a different approach. If cultural organizations fail to acknowledge the differences among the various sources, they risk losing one source or another. When they rely too much on the government, private-sector gifts will dry up. Similarly, when market forces take over, government may see cause for withdrawing its support. Despite the need for different approaches for different sources, the cultural manager must remain focused on the quality of the art produced.[26]

The Economist has predicted that another golden age of philanthropy and charitable giving is about to occur not only in the United States but around the world. There are some major differences in philanthropy and in attitudes towards giving in North America and Europe. Tax incentives are far more generous in the United States than in European countries. Many Europeans are uneasy about publicly contributing money and are not sure how much to give. Many European charities, especially in the area of culture and the arts, are just getting started in terms of fundraising. For the first time, some governments in Europe are launching campaigns to encourage citizens to give to private philanthropic organizations. Fundraising has become an integral part of cultural institutions, and governments are learning how to encourage it.

So what does the future look like for the arts and for arts management? It will vary from country to country and continent to continent, but cultural funding will come from diverse sources and managers will need to know how to present their institutions and artists in different ways for different funders.

Because of the realities of escalating costs and increasingly segmented publics, arts organizations will be re-examining their missions. Some will pursue local funding more aggressively to sustain artistic quality and their primary mission; others will go for the blockbuster exhibition or the smash hit performance. Still others will survive by keeping costs down and programming for their immediate community using local talent and voluntary staff.

None of these possibilities is better or worse than the others. What they suggest is that the future will require not only diverse sources of funding but also diverse organizational and management structures. Finally, though, culture and the arts will continue to be central to living in a civil society.

Notes

1. Davydd J. Greenwood, "Culture by the Pound: An Anthropological Perspective on Tourism as Cul tural Commodization," *in Hosts and Guests: The Anthropology of Tourism*, ed. Valene L. Smith (Philadelphia: University of Pennsylvania Press, 1989), p. 171–186; accessed August 31, 2007, from http://www.questia.com/PM.qst?a=o&d=43089170#.
2. See http://www.operaam.org/ and http://tcg.org/.
3. W. J. Baumol and W.G. Bowen, *Performing Arts – The Economic Dilemma* (Cambridge, MA: MIT Press, 1967).
4. T. Cowen, *In Praise of Commercial Culture* (Cambridge, MA: Harvard University Press, 2000), p. 1–14; accessed August 31, 2007, from http://www.questia.com/PM.qst?a=o&d=94388478.
5. A. Klamer, L. Petrova, A. Mignosa and Stichtung Economie and Cultuur, "Financing the Arts and Culture in the European Union" (*European Parliament*, November 2006); accessed August 31, 2007, from http://www.helgatrupel.de/fileadmin/user_upload/PDF-Datein/Financing_the_arts_and_culture_in_the_European_Union.pdf.
6. Ibid.
7. Ibid.
8. Ibid.
9. S. Toepler, "The Role and Changing Face of Non-market Provision of Culture in the United States" (*Museum International,* Vol. 58, No. 4, 2006), p. 55–63; accessed September 3, 2007, from http:/www.blackwell-synergy.com/doi/pdf/10.1111/j.1468-0033.2006.00583.x.
10. The material here is based in part on information developed by Fundwell.Com: http://fundwell com/interactive/module_viewer.php?page_number=2&chapter_number=9&module_id=23.
11. Louise Jury, "Private Cash for the Arts Hits Record High but Donations from Big Businesses Stall" (*The Independent*, December 12, 2006); accessed September 4, 2007, from http://news.independent.co.uk/uk/this_britain/article2067617.ece.
12. Blackbaud Whitepaper, *Improving Attendance Rates to Increase Donations at Arts and Cultural Organization*; accessed September 4, 2007, from http://www.blackbaud.co.uk/files/bbe/whitepapers/7-07_BBE_ArtsAndCultural_WhitePaper_Web.pdf.
13. For more information, see *Giving USA 2006* (Glenview, IL: Giving USA Foundation).
14. See IEG Network: http://www.sponsorship.com/forum/glossary.html.
15. Ibid.
16. http://www.coneinc.com/Pages/pr_30.
17. Business2Arts. "National Arts Sponsorship Survey: Full Report" (2006); available from http:/www.business2arts.ie/nationalarts6.pdf (accessed September 3, 2007).
18. IEG Sponsorship Report. "Mixed Bag: Sponsorship Spending to Surge in 2008, but Not Everyone Will Benefit," Vol. 26, No. 4 (December 24, 2007), pp. 1 and 4.
19. See Romesch Ratnesar, "Doing Well by Doing Good" (*New Republic*, January 6, 1997), p. 4.
20. "Sponsorship Outlook '99" Sponsorship Strategies, 1998. http://www.amarach.com/amarach/sponsreports.htm.
21. *1999 Cone/Roper Cause Related Trends Report: The Evolution of Cause Branding.* http://www.roper.com/news/content/news115.htm. See also http://www.coneinc.com/Pages/pr_33.html, which summarizes a 2004 study showing that consumers acceptance of such sponsorship has increased.
22. V. Fisher and R. Brouillet, "Les commandites : la pub de demain" (Montreal: Editions Saint-Martin, 1990), p. 15.
23. IEG Network, http://www.sponsorship.com/forum/FAQ.html.
24. "How Social Responsibility Gets Results" (Welland, ON: Tri-Media Marketing & Publicity); http://www.tri-media.com/info-idx.html.
25. Norah McClintock, "Why You Need a Sponsorship Policy and How to Get One" (*Front and Centre*, Vol. 3, No. 5, 1996), p. 12–13; distributed by Canadian Centre for Philanthropy: http://www.ccp.ca/information/documents/fc102.htm.
26. Klamer et al., op cit.; accessed April 17, 2008.

PLAN

Chapter 5

Planning for International Touring in the Performing Arts

By Jennifer Radbourne

OBJECTIVES

- Identify the marketing rationale for taking the performing arts outside the local and national market.
- Evaluate principles for effective international touring.
- Describe the steps in developing and planning an international tour for a performing arts company or product.
- Consider theories and issues of international market entry and regulation, foreign political environments, cultural diversity and product life cycle and distribution in relation to performing arts touring.
- Understand the logistics, promotion, artist management and sponsorship strategies for international performing arts touring or product distribution.

INTRODUCTION

Marketing planning takes on an entirely new dimension in the international environment. When a company decides it is ready to tour internationally, the marketing plan requires considerable research and resourcing to facilitate the logistics of uplifting artists, costumes, equipment, sets, technicians and a touring manager, to produce a work that achieves established goals of new markets, new benchmarks, new reputation and new partners. Globalization of the media has given audiences across the world access to the performing arts of all nations. For many years live performances of international companies have been core components of international festivals. In the 21st century there is an imperative for companies to position themselves in a global market, where national identity and social responsibility take on new meaning. This chapter outlines the marketing case and the planning required for effective international performing arts touring.

5.1 GLOBALIZATION AND INTERNATIONALIZATION

Internationalization involves reaching out to other nations and exchanging ideas and content to deepen understanding, while globalization signifies seeing oneself and one's nation as part of the global community, not separate, not representing an identity to be showcased for acceptance. The dangers of globalization lie in homogeneity, suppressing an identity for the good of the whole. The values of globalization lie in the departure from egocentric practices and a vision and realization of goals that are accepted in all regions. Globalization is inevitable and irreversible. It has become a major force driving creative production, and the arts have been redefined as a result of this and are now clearly linked to a global market.

The transformation that has occurred as money, goods, people, technology and ideas move across borders at an accelerated pace has created opportunities and problems for the performing arts and artists. The "liveness" of the performing arts, which ensures that the performance is a cultural exchange between actors, musicians or dancers with their audience, is challenged by relocation of the production and its cultural meaning. However, from an economic perspective, globalization has created vast new markets, new knowledge, a new interdependence and new wealth. The problem of homogenizing or massifying culture is significant, particularly when the political and economic drivers in the arts are export, touring, international exposure and market development.

For some companies, international touring is a core marketing strategy to achieve international reputation and meet audience demand. In preparing for an international tour, the company must discuss the impact of internationalization on its national or local position, on local audiences, and on its artists and programming.

5.2 INTERNATIONAL TOURING

Performing arts organizations have been attracted to international touring in the same way that corporations take strategic advantage of foreign markets for cost-effective production, maximizing global communications, meeting universal consumer needs, investing in the future and, particularly, managing competition from foreign companies trading in the home country.

International touring is the activity that takes place when an arts organization travels away from its home country to present performance works or exhibitions for international audiences in other countries. The tour may be to one or more countries, and it may be produced by the host country or organization, by a promoter or by the organization itself. In general, the purpose of international touring is to increase market size and awareness and enhance artistic reputation.

Just as large corporations operate globally, many small companies have found success, particularly through the Internet, by broadening their market. International business is not confined to large multinational corporations. When a performing arts company decides to tour, it is considered unnecessary to conduct research into theories of international business. These theories include an analysis of country differences, cross-border trade and investment, the foreign exchange market, strategy and globalization, foreign market entry, global markets and product development, international pricing strategies, and global human resources management. This chapter includes reference to all of these theories as applications in the principles of international touring.

The **marketing rationale** for international touring will drive the international touring marketing plan. This rationale may be an existing contact or invitation to tour, or to demonstrate competitive positioning and the desire for an international reputation, or to use grant money for touring from the government or a sponsor, or even a strong entrepreneurial leader with an international vision. "It is important for the company to examine its motivation for touring abroad. The reasons for touring will influence the frequency and the specific markets targeted and the venues and festivals within those markets. As in any other area of the company's development, clarity of purpose underpins success."[1] A performing arts company may pursue international touring for any (or all) of the following reasons:

- to establish an international brand identity
- to expand audiences
- to promote national identity through culture and the arts
- to raise the profile of the company
- to fulfil artistic development objectives
- to extend the product life cycle of the company's shows.[2]

Performing arts touring presents opportunities and challenges. It showcases the company, the art form and the nation to audiences in another country. It enhances the reputation of the artists and the company, and it leads to sustainability through joint ventures and increased audience awareness. The main avenue for international touring is involvement in international arts festivals.

Arts festivals represent an interesting dimension of both global consumption and nationalistic promotion. Festival performances are packaged into an event that includes disparate experiences and a range of audience choices. According to Bourdeau, De Coster and Paradis, it is as important to assess the festival as a conglomerate product as it is to assess the suitability of the individual performance products within festivals. "Festival organisers must therefore anticipate the expectations of both tourists and residents... They must plan the recreational, educational and touristic aspects of the festival...while also taking into account such factors as the revenues that it will generate...and the community spirit in which the event will evolve." It is important to plan how the touring performance makers might interact with the peripheral services offered at festival events and how these services can be incorporated into the distribution channel of festival product.[3] The festival creates a cultist following, despite the fragmented form, making the whole experience difficult to judge. Whether or not the content is nationalistic, festivals have international appeal. They tend to be vibrant and exciting and can change the life of their particular environment during the run of the festival, transforming a city and providing a communal experience. Robyn Archer, an artistic director of the Melbourne and Adelaide arts festivals in Australia, says that festivals have introduced many artists and works to audiences who would otherwise not have the opportunity to see them. Thus festivals create more sophisticated audiences.[4] This participatory experience in live performance is the key, for it transcends exposure to international arts through television, other media and film.

The outcomes for festival consumers include an ability to compare countries and cultural groups by benchmarking their nation against a standard; increased understanding of the nationalistic dimensions of identity and branding of a company or country; appreciation of excellence, as artists develop with increased peer exchange and exposure; and subsidized prices, as governments and sponsors invest in festivals to achieve exposure and market share.

Many performing arts companies and artists take advantage of consumer desire for live performance and tour the same performance to numerous festivals across countries or continents. This reduces costs, exposes the company to a very wide market, and simplifies repertoire and programming. The result can be homogeneity of festival programs – that is, globalized performing arts.

International performing arts festivals are sites of globalization where local cultural product meets and becomes global cultural product. Figure 5.1 illustrates the interaction among local, international and global performing arts products. Three well-known international arts festivals have been chosen to illustrate the balance of national/local and international content. These are the Hong Kong, Adelaide and Edinburgh festivals. The figures for all festivals include associated events such as artists' talks, exhibitions, and writers' and music festivals held in conjunction with the arts festival. International product in the Edinburgh Festival includes all content other than Scottish content. National/local product in the Hong Kong Festival includes content from China and the Hong Kong Special Administrative Region. National/local product in the Adelaide Festival includes all Australian content.

Figure 5.1

National/Local versus International Product at Performing Arts Festivals (expressed as a percentage)

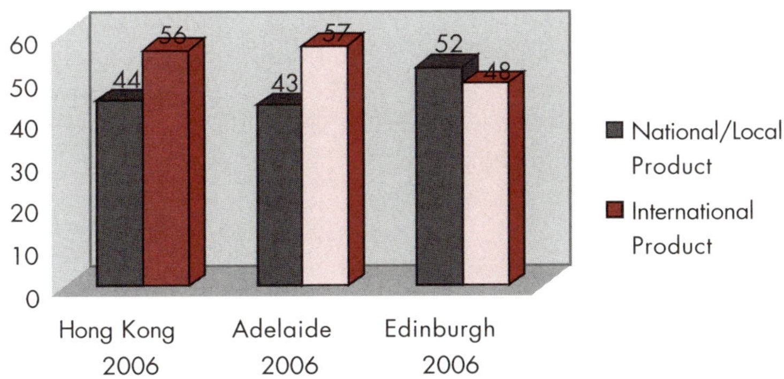

Of the three festivals, Edinburgh is the only one with a majority of local/national product. The Adelaide Arts Festival is designed to be international, with a strong emphasis on international product (with the largest margin between national and international product of the three festivals). In the case of Hong Kong, the emphasis on international product as the drawing card is similar but the margin is smaller.

While international festival participation and international touring are integral to the programs of many diverse performing arts companies, touring can mean very different things to an established company and a small, new company. Small companies often achieve success by viewing the home country as an international market, and their link to globalization. Performing at one of the leading arts festivals in the home country provides an opportunity to be seen by promoters from abroad and offer to tour overseas. Thus the international touring marketing plan often begins years in advance, with domestic touring and participation in international festivals at home.

5.3 THE PRINCIPLES OF INTERNATIONAL TOURING

The following principles provide a critical checklist for any performing arts company planning to tour its product internationally. Each principle is followed by a description of the key issues and examples of application.

5.3.1 International Touring Must Be a Valid Activity Designed to Achieve the Company's Mission and Strategic Development Goals

In order to determine the "validity" of international touring for the company, management should conduct an audit of the company's readiness to tour internationally. In some cases, experience with touring domestically or nationally may be sufficient to have established expertise. Determination of export readiness involves an honest assessment of the human and financial resources for touring, the suitability or adaptability of the product or program, knowledge of the international market or the host organization's level of support in finding an audience, and capacity to embrace the cultural differences that will be experienced.

The criteria for "export readiness" are:

- available finances
- overseas market knowledge and contacts
- recognition in local market
- preparedness to tailor the product
- export market planning
- management resources
- willingness to deal cross-culturally[5]

This audit should also address the company's mission and government policy regarding export, trade and travel with respect to this country. If the company's mission is confined to a particular community that is specifically not international, then international touring must be considered by the board in governance deliberations. The mission may in fact be strengthened by international benchmarking or positioning. The strategic plan for the forthcoming years should also be examined, to determine whether international touring will inhibit other plans or projects.

International touring helps performers to enhance their skills and career opportunities, thus benefiting the company indirectly. "International touring offers opportunities and challenges for the artists working within the company, through: contact with their peers abroad; exchange of ideas or skills development; and opening opportunities for joint work. Relationships developed with individual peer artists may stimulate joint ventures or touring opportunities – indeed the maintenance of the professional relationship may be an end in itself."[6]

The core artistic objectives of the company should be translated into marketing objectives for artistic excellence and audience development. International touring should validate these and achieve international benchmarking and artistic development.

5.3.2 A Set of Measurable and Strategic Objectives for International Touring Should Be Established and Evaluated by the Company

International touring should be proactive rather than reactive – that is, it should be a planned activity that relates directly to the strategic objectives of the company. While some companies may be enticed to tour without any planning or experience but hosted and managed by a promoter, most touring should come out of development goals for audiences and positioning. It is important to undertake a realistic analysis of the company, the product and the markets: "Competitive strategy decisions are developed in terms of a company's internal strengths and weaknesses relative to external opportunities and threats."[7] While the company's strategic plan may not have an international touring strategy, it will have a strategy for new audiences/markets or artistic research and development. It is this part of the strategic plan that is developed into a strategic plan for touring. An international tour requires a specifically constructed strategic plan, with its own mission, situation analysis, rationale for growth, objectives, marketing strategies, budgets and action plan. Such a separate plan ensures that resources and performance measures are set for the tour and evaluated separately.

The rationale for touring guides the formulation of the objectives of the strategic plan. Such objectives are developed in the context of a situation analysis based on predetermined criteria, similar to a "control system," the characteristics of which would include accuracy (expected performances vs. actual performances), timeliness (year and stage of development), objectivity (perceptual biases in international contexts), acceptability (stakeholder expectations and satisfaction), clarity (of purpose), cost-effectiveness (budget) and relevance (reporting on mission).[8] These controls are typically used by corporations planning international business. Their application to the performing arts requires arts managers to be very focused and objective in developing touring strategies.

In the best case, the goal should be to achieve "competitive advantage" by concentrating on innovation, improvement, change and sustainability of product.[9]

Typical international touring strategies may be:

- to increase audiences and the life cycle of the company's repertoire
- to provide opportunities for artistic development and innovation
- to promote the profile of the company as an internationally recognized brand
- to ensure that the international tour is cost-effective

Each of these strategic objectives can be measured in terms of audience attendance, number of performances, critical reviews and awards, brand awareness and company reputation, and costs against revenue.

5.3.3 A Specific Marketing Plan Tailored to International Touring and Supportive of the Organization's Mission Must Be Constructed

An international tour is more than a marketing strategy for audience development. While performing arts companies are always searching for new audiences, shareholders and "funders of the arts want to see targets for growth and strategies for bringing more people and widening access 'for the many, not the few.' In fact, they want evidence of goals and strategies, they want to see marketing plans."[10] The marketing plan for international touring should begin with a specific mission statement for touring that is based on the mission statement for the company. The focus of the plan should be the professional and promotional development of the company and identified touring goals. There are four factors to consider in the international touring marketing plan. These are the product life cycle for the company's current program or repertoire, the need and capacity for market research in the international environment, the market entry conditions in the new environment, and the proposed market penetration and audience development strategies.

A full marketing analysis includes a detailed analysis of the estimated **product life cycle**: Is it in the infancy, growth, maturity or decline stage? This will affect marketing and strategic planning. In the infancy stage, emphasis should be placed on visibility and awareness (to create demand), promotion, risk-taking, initial cash drain and scarcity of competitors. In the growth stage, emphasis should be placed on increasing demand and competition for sales, high advertising and promotional expenditure, where markets would be growing faster than the economy. In the maturity stage, emphasis should be placed on market research, product testing, cost controls, stability of competition and possible mergers. In the decline stage, emphasis should be placed on minimizing change in market share, minimal promotion, competition through price-cutting, stability of location and customer base.[11] For the touring performance or program, therefore, the company should use this analysis to support a strategy for a new audience for an existing program, or using the tour to present a new work representing the company and the nation. Festivals are excellent opportunities to try new works for new audiences prepared to take artistic risks. Often, an international tour is a product life-extension strategy, taking a well-known product away from its traditional audience and giving it to a new audience before decline sets in. It is also possible that the program is in different stages in its life cycle in different international markets, so that the mature program in the home country is in infancy or growth in the international market, thus requiring different investment and promotion.

This internal market analysis must be balanced by as much external **market research** of the international market as possible. This research could encompass the usual surveys and interviews, experimental studies, consumer panels, customer profiles, and the effectiveness of specific promotional tools.[12] Most performing arts companies will not have the capacity to conduct market research in an international environment and must rely on previous tours or the tours of other companies in their area. In many cases the data will be available from secondary sources or directly from the host organization or promoter. But to ensure that the right program is selected and the company can sustain the costs of touring, research is critical – to dentify the target markets and their consumer behaviour for international programs. Also needed is a thorough assessment of the "attractiveness of markets," measuring potential benefits against potential risks. Benefits would include monetary benefits, the size of the market (audience), the present wealth of consumers and the likely future wealth of consumers. Risks would include political risk (the likelihood of a major policy change in the business or arts environment), economic risks (a change in the buying power of audiences or sponsors) and legal risks (a breach of contract or expropriation of property rights).[13]

The marketing team can now focus on **market entry** choices. For example, different markets require different financial models, such as licensing agreements, franchising, contracts, joint ventures or strategic alliances. Market entry decisions will also be affected by the compatibility of product and market(s), attractiveness of the location, availability of performance spaces, availability of audiences and the perceived risks.[14] It would be unwise to take a production with complex use of the English language to a non-English-speaking country, or to take a performance requiring a special dance floor and specific stage heights for leaps and jumps to a stage that has low ceiling space, or to tour to a country where political turbulence prevents audiences from travelling to public venues after a certain hour. Risks can be reduced by balancing "marketing standardization," such as "international market grouping," with "marketing adaptation," such as "limited adaptation" of distribution, promotion and product for different target markets.[15] Research means that the product may be more acceptable grouped in a festival, or can be adapted for the international market and even developed specifically for touring without compromising or "dumbing down" the program. For example, many companies choose to tour a show with a small cast and limited set requirements. Communication strategies should be designed well in advance of the tour, taking into account the length and variety of distribution channels and the "push-pull" mix of the promotional tools in the host country. International representation and the availability of media need to be considered, to maximize promotion opportunities. Some companies consider it critical to make professional contact or use personal international relationships before entering a new market.[16]

When planning for market entry and **penetration**, companies need to research the characteristics of specific markets before deciding where to focus their activities. Penetration is the strategy used to make an impact on an audience, either by mass promotion or by direct promotion to a selected audience who will engage with this performance. Questions should be asked about the long-term potential for development in this market. Has any interest been shown in the company's work? Is there a reasonable degree of economic and political stability? How much competition is the company likely to face in establishing a presence in this market? How receptive to the company's work are audiences and promoters likely to be?[17] The marketing plan distribution channels should consider the relationship between certain target markets, particularly as this affects the promotion of the product and its life cycle. For example, "Some companies may tour to one market partly to provide entry into another market. London and New York are regarded as discriminating classical music markets, so for classical music ensembles the prestige of having performed in London or New York may assist in penetrating the Japanese or other markets."[18] Similarly, festivals can represent a major distribution channel of the product to target markets.

Audience development should be a key component of the marketing plan for international touring: "In addition to advance scheduling, the company which wishes to maintain a profile touring abroad needs to make a regular commitment in order to build its presence in, and knowledge of, a chosen market. Audience development depends upon such commitment. Return visits provide the opportunity to refine and consolidate the touring circuit, to respond to offers which arise on tour, and to improve the company's touring logistics as it gains familiarity with the technical, travel, administrative and other aspects related to entering the target markets."[19]

The international touring marketing plan must demonstrate an understanding of the company's internationalization strategy. It is vital to understand the branding of the home country in international arts markets, in order to evaluate whether the performing arts product can penetrate highly discerning markets like Japan[20] or whether national identity takes precedence over innovation and increased audiences.

5.3.4 A Detailed Financial Model Should Be Adopted for International Touring, Independent of the Organization's Domestic Financial Plan

In marketing terms, the company should start its financial modelling for international touring with an understanding of the market rate for the work so that the company is not undersold.[21] The financial model should be based on the company's pricing strategy in the domestic market. Components of the pricing strategy may include "predatory pricing," "multipoint pricing" and "experience curve pricing," depending on the size of the company and the markets it is considering entering.[22] Market research in the domestic market enables the company to offer discounts to patrons on the basis of age or employment status or as part of an introductory offer or a package. By pricing below other companies, the company may be able to increase its market share. Multipoint pricing is used for subscription packages. In the international market, most of these strategies are not possible for individual companies but may be used by festival organizers. The excitement of an invitation to tour or the allure of an overseas market must not cloud the normal accountabilities and responsibilities for the financial performance of the company.

The financial models for performing arts companies are diverse. They include joint venture, strategic alliance, sponsorship or philanthropy, government funding, self-funding, fully funded invitation and complete purchase of the performance by a promoter or international producer.

Joint ventures or strategic alliances with companies in foreign markets, or with equivalent performing arts companies in foreign markets, or with touring/booking agents or festivals, represent shared financial risk in touring. "Countertrade" possibilities – that is, exchange agreements between arts festivals – also offer security and shared financial responsibility. **Sponsorship,** both individual and corporate, provides a marketing or business benefit to a company trading in the selected overseas market. The company can use the arts performance to host business partners and promote shared values. Corporate sponsorship should be tailored to the size of the performing arts company and the breadth of its audience, because servicing a sponsorship agreement can take more resourcing than the company can deliver. In some instances, a charity or foundation may provide **venture philanthropy** for the purpose of capacity-building or delivering an arts program to the overseas community in a gesture of social responsibility for increased awareness and understanding. **Government funding** for the arts, particularly that tailored to international touring or the export of national products, can be sourced both through arts budgets and through funds allocated to the export of national cultural products. Most governments have an Export Market Development Grant Scheme. This may be useful for arts organizations embarking on a concerted export effort.

Performing arts markets can be an effective stimulant to international touring: "A significant number of Australian performing arts organisations have benefited from a presence at the Australia Council's Australian Performing Arts Markets, where they have been seen and picked up by international presenters and promoters. There are also performing arts markets in other countries, for example in Japan (Tokyo) and Canada (CINARS in Montreal)."[23] In some instances the promoter or producer will buy the company's program outright. This involves payment for service and the company must ensure that the agreement does not infringe on moral rights.

The **self-funded tour** involves considerable financial risk on the part of the performing arts company. It usually results from a previously successful tour and direct marketing to an established audience, or from the fact that the company has excellent market data and sufficient experience to ensure a commercial return. A **fully funded invitation** is not uncommon in the global arts environment, where governments and international festivals invest heavily in building their profile through international partners and exposure. It is a benchmarking and identity development strategy. The performing arts company should scrutinize the invitation and ensure that all requirements are included in the agreement that follows it.

5.3.5 Market-Specific Operational Logistics Must Be Carefully Planned, with a Particular Emphasis on the Cultural, Political and Legal Contexts of the Touring Destination

While logistics relate to the detailed operation of the tour, there are some "big picture" factors that will influence the logistics of delivery and therefore must be considered. For example, in international trade activity there must be an understanding of the political relationships between the home country and the country identified for the tour, the cultural preferences in this country, its geographic proximity to the home country, the currency and exchange rate, and the potential for inflation to change the costs of touring in a year or two. International law may affect intellectual property rights or employment contracts. All these factors will have an impact on the program selection, sources of funding, touring costs, government support and existing contracts for artists.

Pre-tour work should include applying for passports and visas or work permits, arranging necessary inoculations for artists, preparing for any immigration procedures, making accommodation arrangements, calculating per diems for company members, arranging for travel and public-liability insurance, forecasting local production costs (local crew, lighting and sound equipment hire) and freight arrangements, making travel arrangements for artists and technical crew, developing a policy on holidays and public holidays overseas, and reckoning mobile phone charges. In the case of performing arts, it is imperative that cultural/linguistic contexts be taken into account when preparing detailed technical specifications for the performance, as translation may be necessary. The home country's embassy in the host country can be very helpful here, giving culturally specific advice and promoting the company or arranging additional performance opportunities.[24]

The budget must be developed following research on tax arrangements, fee structures and per diems. Other companies who have entered the same market may be able to offer advice. The budget should cover all income and expenditure projected for the project, and should be separate from the company's overall budget. The company may need to take care of sponsors' needs, such as agreed credits on publicity (including pre-sighting of promotional material prior to touring), ticket allocation for guests and corporate hospitality benefits. These should also be included in the budget and the logistics plan, showing what is to happen, and when.

A pre-tour visit (six months in advance), to make contact with local technical staff, assess the site and performance spaces, and give venue organizers detailed technical specifications, will reduce risk and facilitate planning. On this visit, media liaison could begin in order to optimize media coverage during the tour. During the tour, the director or manager should be available for interviews and future promoters should be invited to performances. It is preferable to provide the publicity material for international markets. "Maintaining a stock of strong visual images for promotional purposes is important."[25] It may help to employ an international company representative to assist with touring. In addition to securing bookings, the representative can provide touring advice, handle some of the tour logistics and relieve the burden on the company's administration. A contact in the target market with strong local knowledge will be able to represent the interests of the company and at the same time manage touring logistics.[26]

5.3.6 A Retrospective Analysis Should Be Used to Assess the Success of International Touring and to Develop Future Touring Plans

As with any marketing plan, monitoring and evaluation are critical. In the post-tour analysis, the touring outcomes must be balanced against the touring objectives. For the company's own future activities, the more detailed the process of debriefing, the better. The evaluation could be triggered by the following questions: What useful contacts were made? What unanticipated issues or problems arose? What aspects of planning or tour logistics should be handled differently next time? Have the performance targets been achieved? Has the company's reputation been enhanced? What was the audience response? How has the company benefited artistically? What should be included in the sponsor or government report on the tour?

Links with contacts in international markets should be maintained, to ensure future touring and input into measuring the success. Prompt financial acquittal is necessary, to determine the cost-effectiveness of the tour as well as to maintain business and sponsor relationships with a high degree of professionalism.

Planning for future touring seasons should be initiated immediately after the tour. The Board can use the evaluation data in strategic development discussions.

5.4 STAGES IN INTERNATIONAL MARKETING PLANNING

These principles must be included in the three-stage marketing plan for an international tour. **Stage 1** is the most complex of the three stages. It involves the pre-tour planning. It begins with the executive, artistic and administrative staff developing a feasibility study or proposal for the tour and presenting this to the Board of Directors for discussion and endorsement.

A **feasibility study** is a study of the desirability and practicality of the proposed tour. It is an opportunity for the company to examine thoroughly what it hopes to achieve and determine whether it is able to do so. Standard assessment criteria in a feasibility study are:

- artistic feasibility (artistic plans, other performing arts providers in the target market, arts funding requirements, the artistic mission)
- technical/physical feasibility (product/program/performance, design features, functionality of technical infrastructure, number and availability of artists, venue specifications)
- financial feasibility (funding, cost of operation and maintenance, cash flow)
- organizational feasibility (touring management strength, corporate structure, support from the Board)

The decision-making at this time should address the preconditions for success, such as organizational preparedness, long-term commitments and plan, stability of the company, personal contacts and relationships overseas, capacity for flexibility and resiliency, and commitment to audience development. When the decision to tour has been made, a marketing plan should be developed to guide the export activities. This plan includes specific international touring marketing objectives, marketing strategies, a budget and an action plan. The action plan contains the logistical details of presenter/promoter research, fee negotiations, market research, sources of funding, sponsor/presenter/host liaison and servicing, venue research, booking and scheduling, transport options, artist accommodation and per diems, visa requirements, local taxes and regulatory controls, insurance, royalties, promotional material, press liaison, props and/or set availability and costume maintenance.

Stage 2 of the touring marketing plan is the tour itself. In this section of the action plan, the performance measures are delivered. The company leaves its home base, travels to the host country, performs, and achieves its objectives, which may be audience numbers, audience response, press coverage, new relationships, artistic development, intercultural development and international exposure. The action plan moves from planning to delivery and measurement.

Stage 3 involves evaluating the performance against target measures. As well as the key performance measures set down in the marketing plan, the company must measure success against the strategic objectives for touring and the financial outcome of the tour. A rigorous analysis of the impact of the tour on the company's positioning is critical to determining the benefits against the mission of the company. A survey of local audiences, sponsors and/or funders, competitors, media and artists will help the Board of Directors and management determine the success of the tour.

The international touring marketing plan will look like the one shown in Figure 5.2.

Figure 5.2
International Touring Marketing Plan

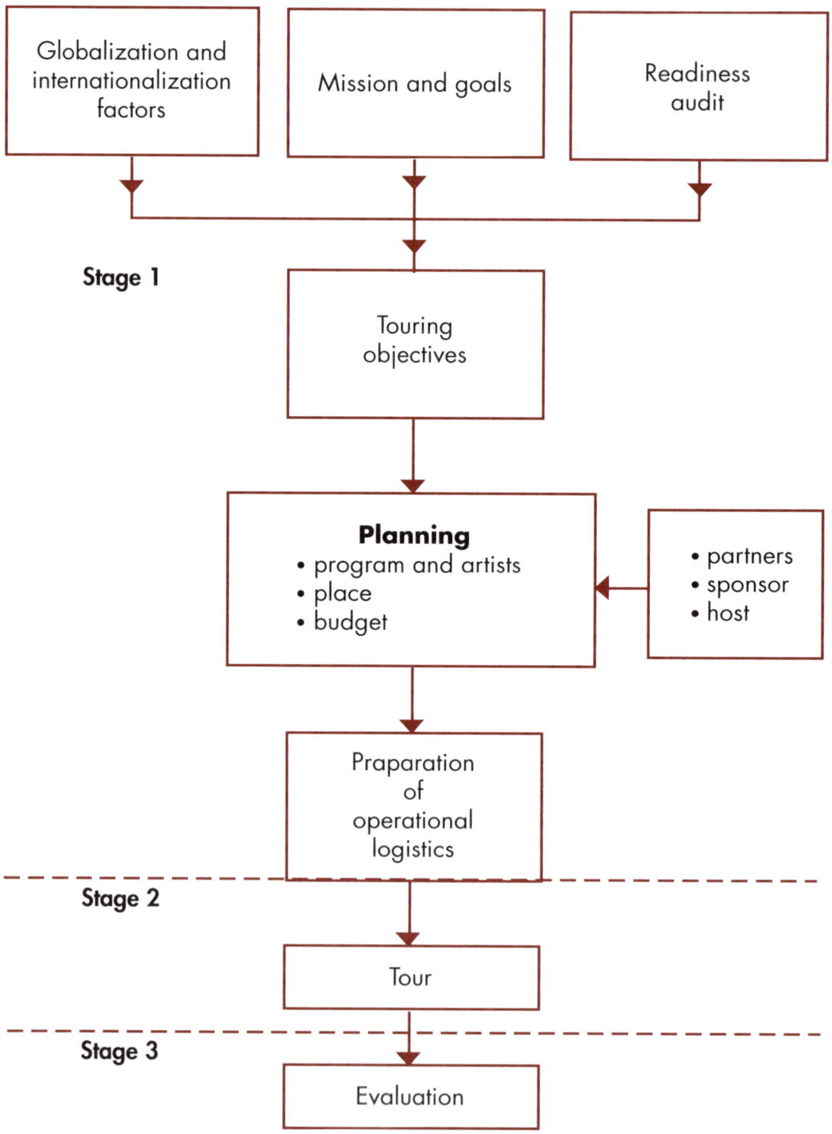

5.5 CASE STUDIES

The three case studies that follow allow the reader to examine the companies and their international touring activity with this planning model in mind. How does the mission provide for touring? What characteristics of the company made it ready to tour? What market research was done? What were the motivations for international touring and what were its objectives? What program was chosen for the overseas country? Why? What financial model was used? How were the touring logistics managed? What was the audience response? How does the company's international touring affect its domestic performances and positioning?

The first case is Cirque du Soleil, a large franchised corporation with a strong social responsibility mission. The second is Chanticleer, a niche company with niche international audiences validating its mission of educational outreach in a cappella music. The third is Strange Fruit, a small group of community artists presenting street theatre and supported more by international festivals than by domestic audiences.

5.6 CASE STUDY: CIRQUE DU SOLEIL

Cirque du Soleil is a global franchised company with a demonstrated focus on corporate citizenship and global social responsibility. It started as street theatre in the early 1980s. In 1982, a 22-year-old fire-eater, Guy Laliberté, along with Gilles Ste-Croix, a stiltwalker, and Daniel Gauthier, who would become the Cirque's president, founded a summer festival in Baie Saint-Paul, Quebec, that featured their friends – stiltwalkers, acrobats and other street entertainers. Their show was a hit, and in 1984, when the provincial government invited festivals to help celebrate the 450th anniversary of Jacques Cartier's voyage to North America, Laliberté and Gauthier, in typical Canadian fashion, applied for and received a grant of $1.5 million from the Quebec government to create Cirque du Soleil.[27] "More than 1,000 of Cirque du Soleil's employees are based in Montreal, where the company is headquartered."[28]

"The rationale and values behind Cirque's social action are grounded in this history where youth, risk, dreams and marginality come together for a better world."[29] One percent of annual revenue is allotted to outreach programs targeting youth in difficulty. The company is committed to corporate sponsorship as one of its revenue sources but also as a measure of its social mission. "As part of our sponsorship philosophy, Cirque du Soleil seeks partnerships with companies that are truly committed to making their names synonymous with the things that make Cirque du Soleil great. These include a commitment to excellence, attention to the people behind the scenes, and a strong involvement in the community, both locally and globally."[30]

A five-year plan developed in 2002 described the company's strategy as continuing to be different from Disney, MGM and other entertainment rivals. Cirque du Soleil styles itself as a pure content provider whose main business is harnessing the creativity of performers, producers and other artists, not owning and operating hotels and other properties. It focuses its attention and finances on training and other activities related to the creative process. Its target audience is adults, not children, with tickets costing approximately $100. The company believes each individual show can be kept alive for up to 15 years before it has to be retired.[31]

"In 1990 Cirque du Soleil went to Europe, in 1992 to Japan, and everywhere the response was awe and delight. Now there are permanent facilities in Las Vegas and at Walt Disney World in Florida, along with offices in Europe and Asia."[32] The permanent establishments in the United States were a response to cultural sensitivity and international logistics: "Improved access to talent was only part of the motivation for granting Las Vegas its own human resources office...the primary purpose was to build an all-American team with an in-depth knowledge of United States labour laws. As Cirque's United States numbers multiplied, the company realized it couldn't depend on its Montreal office to understand the complexities of the Equal Employment Opportunity Commission's rules, or the peculiar cultural sensitivities of an American work environment."[33]

All around the world, performances are sold out. Audiences marvel at the skill of the artists and the sophistication of the artists and the ideas. Cirque du Soleil operates as a business corporation, investing massively in promotion and technical infrastructure and in making a lasting impression on audiences. The company tours with educational resources for its young artists and its own health-care team. It uses local labour for ushering, cleaning and refreshments. The domestic impact of this level of international touring is continued funding from the Quebec and Montreal governments, which benefit from the international profile despite causing tension among not-for-profit arts companies in that province and city.

The 2007 program included productions by four companies in Dubai, Atlanta, London, Auckland, Barcelona, Houston, Canberra, Melbourne, Colombo, Saint Denis, Gijon, Adelaide, Perth and, in Canada, Montreal and Toronto.

5.7 CASE STUDY: CHANTICLEER

Chanticleer is a 12-man a cappella choral ensemble based in San Francisco's Bay Area, where it presents a season of concerts annually. This is a unique company in a niche market whose objectives are to entertain and educate. It hosts education and community outreach programs locally. The education outreach falls under the Singing in Schools program, established in 1986. This supports the artistic vision of the company as well as its strategic marketing plan, in developing a strong local brand and a sustained product life cycle. These programs are targeted at schoolchildren but include adults as well.

Chanticleer was founded in 1978 by Louis Botto and originally had nine volunteer members. The choir has been employed full time since 1991. In 1990 it was "the only full-time professional a cappella men's chorus in the United States," according to Joseph McLellan of the *Washington Post*.[34] It is now sustained by more than 80 annual concerts in the United States alone. The ensemble has become a staple of the Bay Area concert season but also has a presence on the East Coast, presenting a series of concerts annually at the Temple of Dendur in the Metropolitan Museum of Art in New York City.

The company began touring internationally in 1984 with a performance at the Josquin Symposium in Belgium. In 2006 it performed at festivals in Japan, France, Austria, Germany and Switzerland. The ensemble's small size and the nature of its product make the selection of international markets relatively simple. The touring destinations are those with an interest in this type of music, hence Europe and Japan. The company has a full-time tour manager on staff, and while it responds to international festival invitations it now has direct contact in the countries visited, providing logistics and direct mail support. The motivations for touring are full employment for the ensemble and fulfilling Chanticleer's mission with respect to outreach and education. In addition to performances, the company records its programs, releasing its 29th CD, *Sound and Spirit,* in 2005. It has been signed with the Warner Classics label since 1994.

According to artistic administrator and director of educational outreach Philip Wilder, Chanticleer's mission is "to sing really good music that some people might call esoteric, from Renaissance works to new music and commissions."[35] The ensemble's mission statement, as it appears on the company's Website, is the following: "Chanticleer is a male vocal ensemble that performs a diverse and innovative repertoire, reaching audiences worldwide through live concerts, electronic media, and education."

The company's general marketing strategy comprises five areas: individual membership, corporate sponsorship, educational outreach, national and international touring, and product distribution through CD recordings. Chanticleer relies heavily on individual contributions, with 40% of contributed funds coming from hundreds of individual donations ranging from $50 to $100,000. Multilevel corporate sponsorship packages are invited for annual contributions ranging from $500 to $15,000. The Chanticleer Endowment Foundation was established in 1996 to manage the company's philanthropic funds, and the Joseph H. Jennings Fund for Music was established in 2004 to "underwrite the costs of commissioning new choral works."[36] To date, Chanticleer has commissioned over 60 compositions in eight to 12 parts from a long list of composers. Because of Chanticleer's extensive national and international touring, new works commissioned by the group will often have 50 performances to general audiences within two years after their premieres, and CD recordings of the pieces exponentially increase those numbers. "Few other choral ensembles can offer this kind of exposure and accessibility to new music."[37]

Between September 2004 and September 2005, Chanticleer toured only nationally, covering 30 states and presenting 82 shows.[38] Between November 2005 and March 2006 it toured internationally, to Japan, France, Holland, Austria, Spain, Estonia, Lithuania and Latvia, giving 19 performances in 19 cities. During this period the company's international touring represented 49% of its total number of performances. The 2007 program was again focused on the United States, including Hawaii, with an emphasis on San Francisco.

5.8 CASE STUDY: STRANGE FRUIT

"Strange Fruit specialises in large-scale, outdoor public performance. The work is a refreshing synthesis of sculpture and performance, with very strong visual and design elements and startling choreography."[39]

This Australian company performs street theatre staged on 4.5-metre poles – a combination of "dance, vaudeville, popular and classical music, circus and sculpture."[40] It was established in 1994 and gave its premiere performance at the Melbourne International Festival in October of that year with *The Field*. This production has now been performed to more than one million people around the world.[41] Since 2003 Strange Fruit has comprised 30 performers who can form into two or three groups upon demand.[42] The company has received some funding from different tiers of government – municipal, state and federal – but does not have ongoing corporate sponsors.

Strange Fruit has twice won the silver trophy at Japan's Daidogei world cup of street theatre, in 1999 and 2000, and it won the Premio del Publico in Valladolid, Spain, in 2003. But while it is acclaimed around the world, the group struggles to secure adequate funding in Australia.[43] The company has found it easier to enter international markets than national ones and has suggested that street performance is not a recognizably Australian product. It believes that international touring supplements a lack of national demand for product and that international markets extend the product life cycle. Strange Fruit has struggled to reorient its marketing back to its home audiences. "The perception is that we are overseas all the time and we want to change that,"[44] stated the group's director, Roderick Poole, in 2003. The strategy has been to design performances for special events, product launches and gala dinners, adding unique entertainment as guests arrive or delivering an engaging performance piece as the highlight of an event. Examples of specially tailored corporate performances include *Airbourne,* a show custom designed for the opening of the new Adelaide airport in 2005, and *Swoon and Spheres,* performed on Brisbane People's Day, December 2, 2006, at the opening of Queensland's new Gallery of Modern Art.

This small performing arts company first toured internationally, to Europe, in 1996 after being picked up at the first Australian performing arts market in Canberra.[45] The company's first European tour covered four countries – France, Belgium, the Netherlands and the United Kingdom – giving a total of 45 performances.[46] "The company has performed in over 150 international festivals in 28 countries over the last six years."[47]

Poole is aware of the cultural differences among the different international audiences, and the performances take this into account. "In Asia, people respond to the element of danger and the circus elements… In Europe, it is the theatre styles of mime and commedia dell'arte."[48] Strange Fruit has tailored its product to suit the widest possible range of audiences: "Every show in our repertoire is an artistic expression of a universal theme, such as love, conflict, birth, death, work and play. All of them are performed on poles and typically play to an audience in the round. They suit a wide variety of environments – indoor and outdoor, day and night, festival and corporate settings. Each show in the repertoire has different technical requirements, such as a height clearance of seven metres, an area space of at least 8x4metres for a two person show and 16x8metres for an eight person show, clear weather conditions for outdoor performances, and quality sound and lighting."[49]

Strange Fruit now has touring representatives in Australia and Asia, based in Melbourne, and in Europe, based in Belgium. "Due to the dialogue-free nature of the work and its ability to speak to people across all divides, the company's repertoire is extremely popular with international audiences and most of the company's work is performed overseas, resulting in the development of tightly synchronised logistics and tour management professionals. At any one time, there can be performances taking place on three different continents."[50] In 2006 Strange Fruit toured to Poland, Britain, Croatia, France, Spain, Germany, Italy, the Netherlands and Belgium, performing 21 shows between June 22 and August 20.

SUMMARY

This chapter has outlined the marketing tasks and planning stages necessary to take a company on an international tour. The three case studies have illustrated three very different motivations and successful methods of international performing arts touring. There are many examples of touring disasters, with companies taking up to 10 years to recover financially from over-ambitious touring plans. The risks for a company in touring are many, from cultural to political to financial, including negative reviews and poor audience reception. However, with a rigorous plan that strategically develops the company's mission and positioning, international touring can be highly rewarding. International touring of live performing arts contributes in a unique way to the "pushing out" of national identity and difference at a time when globalization is facilitating massification of the performing arts.

Notes

1. Australia Council and Arts Victoria, *Let's Tour! A Quick Guide to Reporting the Australian Performing Arts* (Victoria: Australia Council, 2000), p. 2.
2. Ibid.
3. L. Bourdeau, L. De Coster and S. Paradis, "Measuring Satisfaction among Festivalgoers: Differences between Tourists and Residents as Visitors to a Music Festival in an Urban Environment" (*International Journal of Arts Management*, Vol. 3, No. 2, 2001), p. 40–41.
4. Gabriela Coslovich, *The Age*, September 30, 2006, A2, p. 19.
5. Australia Council and Arts Victoria, op. cit., p. 5.
6. Ibid., p. 2.
7. Betty Jane Punnett and David A. Ricks, *International Business*, 2nd ed. (Cambridge: Basil Blackwell, 1997), p. 329.
8. Ibid., p. 408–411.
9. Ibid., p. 331–332.
10. F. Kerrigan and P. Fraser, *Arts Marketing* (Amsterdam: Elsevier Butterworth-Heinemann, 2004), p. 193–194.
11. Punnett and Ricks, op. cit., p. 214–216.
12. Ibid., p. 337–344.
13. Charles W. L. Hill, *Global Business*, 2nd ed. (New York: McGraw-Hill, 2003), p. 74–76.
14. Punnett and Ricks, op. cit., p. 247–260.
15. Ibid., p. 331–333.
16. Australia Council and Arts Victoria, op. cit., p. 3.
17. Ibid.
18. Ibid., p. 2.
19. Ibid., p. 4–5.
20. Ibid., p. 33.
21. Ibid., p. 18.
22. Punnett and Ricks, op. cit., p. 412–414.

23. Australia Council and Arts Victoria, op. cit., p. 8.
24. Ibid., p. 5.
25. Ibid., p. 7.
26. Australia Council and Arts Victoria, op. cit., p. 6.
27. "Cirque Du Soleil," *Maclean's*, Vol. 113, No. 36, 2000, p. 45.
28. T. Peterson, "Cirque Du Soleil's Expanding Big Top" (*Business Week Online*, 2002).
29. http://www.cirquedusoleil.com/CirqueDuSoleil/company/socialaction/default.htm.
30. Ibid.
31. Ibid.
32. "Cirque Du Soleil," op. cit., p. 45.
33. Cindy Waxer, "Life's a Balancing Act for Cirque Du Soleil's Human Resources Troupe" (*Workforce Management*, Vol. 84, No. 1, 2005), p. 52.
34. Joseph McLellan, "Getting Small: The Little Big Music of Orpheus and Chanticleer" (*Washington Post*, February 4, 1990), p. g1.
35. Steve Smith, "The Classical Score" (*Billboard*, Vol. 113, No. 49, 2001), p. 27.
36. http://www.chanticleer.org/contact.cfm.
37. Ibid.
38. Chanticleer 2004/05 Tour Schedule, provided by Chanticleer.
39. Australia Council, *In Repertoire: A Guide to Australian Contemporary Performance* (Sydney: Author, 2001).
40. "Stars of Pole Position" (*Leader*, November 13, 2000), p. 6.
41. Rebekah Cavanah, "'Best' Festival Exits Stage Left" (*Northern Territory News*, August 29, 2006), p. 8.
42. Robin Usher, "Performance Group Branches Out" (*The Age*, March 28, 2003), p. 4.
43. Ibid.
44. Ibid.
45. Ibid.
46. Australia Council and Arts Victoria, op. cit., p. 36.
47. Australia Council, op. cit.
48. Usher, op. cit., p. 4.
49. http://www.strangefruit.net/au.
50. Ibid.

References

Australia Council. 2001. *In Repertoire: A Guide to Australian Contemporary Performance.* Sydney: Author.

Australia Council and Arts Victoria. 2000. *Let's Tour! A Quick Guide to Exporting the Australian Performing Arts.* Victoria: Author.

Bourdeau, L., L. De Coster and S. Paradis. 2001. "Measuring Satisfaction among Festivalgoers: Differences between Tourists and Residents as Visitors to a Music Festival in an Urban Environment." *International Journal of Arts Management*, Vol. 3, No. 2, p. 40–50.

Cavanah, R. 2006. "'Best' Festival Exits Stage Left." *Northern Territory News,* August 29, p. 8.

Chanticleer Website: http://www.chanticleer.org.

Cirque Du Soleil Website: http://www.cirquedusoleil.com.

"Cirque Du Soleil." *Maclean's*, Vol. 113, No. 36, 2000, p. 45.

Clark, C. 2000. "Balancing Act That Really Takes Off – Sydney Live." *Daily Telegraph*, September 18, p. 85.

Coslovich, G. 2006. *The Age*, September 30, A2, p. 19.

Hill, C.W.L. 2003. *Global Business*, 2nd ed. New York: McGraw-Hill.

Kerrigan, F., P. Fraser and M. Özbilgin. 2004. *Arts Marketing*. Amsterdam: Elsevier Butterworth-Heinemann.

McLellan, J. 1990. "Getting Small: The Little Big Music of Orpheus and Chanticleer." *Washington Post*, February 4, p. g1.

Peterson, T. 2002. "Cirque du Soleil's Expanding Big Top." *Business Week Online*.

Punnett, B. J., and D. A. Ricks. 1997. *International Business*, 2nd ed. Cambridge: Basil Blackwell.

Shagan, R. 1996. *Booking and Tour Management for the Performing Arts* [revised and expanded edition of the book formerly titled *The Road Show*]. New York: Allworth.

Smith, A. W. 2002. "San Francisco's Chanticleer: An American Tale of A Capella Perfection and 'Policy Governance'." *International Journal of Arts Management*, Vol. 5, No. 1, p. 74–83.

Smith, S. 2001. "The Classical Score." *Billboard*, Vol. 113, No. 49, p. 27.

"Stars of Pole Position." 2000. *Leader*, November 13, p. 6.

Usher, R. 2003. "Performance Group Branches Out." *The Age*, March 28, p. 4.

Waxer, C. 2005. "Life's a Balancing Act for Cirque Du Soleil's Human Resources Troupe." *Workforce Management*, Vol. 84, No. 1, p. 52–53.

"Your Own Private Cirque Du Soleil." 2004. *Corporate Meetings and Incentives*, Vol. 23, No. 2, p. 24.

PLAN

Chapter 6

International Marketing for Cultural Industries (Film, Publishing, Recording)

By Johanne Brunet

OBJECTIVES

- Understand some of the international trends in the marketing of cultural industries
- Understand the importance of market research in an international activity
- Develop an understanding of the elements of risks versus control in international activities
- Develop an overall understanding of the major issues in international marketing

INTRODUCTION

Industries, firms, lifestyles and consumer tastes have become international and global, and cultural products and the cultural industries are no exception. Whether for economic reasons, because of openness to other cultures or simply because of a basic curiosity, the internationalization and globalization of the cultural industries is a reality.

Consumers want to discover other cultures and their artistic manifestations, while cultural producers want to sell their products internationally and export their culture. This chapter covers the many ways in which cultural products can be exported and the many factors that must be considered before any export activity can take place.

First we will explore international and global trends within different cultural industries. Then we will look at the incentives and barriers to international activity. Finally, we will review the importance of evaluating the target markets and revisit the marketing mix in the context of internationalization.

6.1 TRENDS IN INTERNATIONALIZATION AND GLOBALIZATION

As we sit in our living rooms watching the latest news prepared by our local television network, CNN, Euronews, the BBC or Al Jazeera, we are in contact with the planet on a daily basis. Around the world, virtually within the same 24-hour period, movie fans are viewing the same Hollywood blockbuster, children are reading the same Harry Potter book and teenagers are listening to the same rap artist. And there is no doubt that the entire planet is watching the latest video clip on YouTube. Meanwhile museums in different countries are coming together to present international exhibitions that will be seen in museums around the world. Artists, actors, singers, musicians and authors view the planet as their playing field.

People are communicating digitally, whether by phone or on the Internet. Communications technology is evolving on a daily basis, making this process easier by the minute.

The export and import of cultural products have become an important part of the commercial balance in different countries. The entertainment industry represents a major export for the United States. A review of the different cultural industries shows that the international market is dominated by the United States, with a handful of multinational conglomerates surrounded by many small enterprises.

6.1.1 Evidence from Different Cultural Sectors

The **film industry** has reached a high level of maturity in terms of internationalization and globalization. It is dominated by the United States. Many reasons explain this situation. First, the sheer size of the American domestic market allows U.S. firms to amortize a large proportion of their initial production costs. The financial contribution of the majors (Warner Bros., Disney, Columbia, Paramount and Fox) gives them access not only to production budgets but also to the all-important promotion budgets and large distribution systems. These American corporations control the theatrical distribution system both domestically and internationally.

Another factor in the U.S. dominance of the film industry is the prominence of the English language internationally. English is the most popular language in the most economically important regions of the world.

The myth of the American Dream still captivates international audiences. Cinematographic production highlights and concentrates on spectacular situations, special effects and appearances.

Table 6.1 gives the total number of films produced in 2006 in the United States, the United Kingdom, France, Italy, Canada, India and all countries of the European Union, and the average production costs. The United States leads the market in terms of both the number of films produced and the

Table 6.1
Films Produced in 2006[1]

Country	Number of Films, (Including Co-productions)	Average Production Cost (US$M)
United States	699 (2005)	65.8
European Union	862	
United Kingdom	78	13.6
France	240	6.6
Italy	86	4.9
Canada	80	4.4
India	1,090[2]	< 1.0

average production cost. While India is a prolific producer, its films have limited international appeal.

Of the top 20 movies in 2006 in terms of international box office, 14 originated in the United States and the other six were co-productions between the United States and the United Kingdom, New Zealand, Germany or Canada.[3]

It is interesting to note that the United States is only a marginal importer of films. The American consumer does not embrace dubbed or subtitled films. Even a foreign accent, such as a British accent, is not well received in the United States. Generally speaking, the American market is not open to foreign products.

The **music industry** has also been a witness to globalization. Not only is the same type of music appreciated globally, but its mode of distribution has been in transition. Digital distribution has had a huge impact on the structure of the music industry. Young people are downloading music from the Internet, global sales of CDs are decreasing and the industry has been very slow to react. Consumer behaviour has changed and the industry has not been adapting. In 2006 Paul McCartney chose not to renew his contract with EMI Records, his label since 1962. He released his new album, *Memory Almost Full,* with Hear Music, a label co-founded by Starbucks; the CD was sold in coffee shops and record stores but mainly on iTunes.[4]

Along with this global aspect of the industry, the independent music sector is flourishing, with young musicians and artists bypassing the traditional distribution and promotion system, using the Web, and developing an international following. For example, the band Simple Plan from Montreal, Canada, has used the Web to develop an international career and following, selling more than seven million copies of its CDs.[5]

The worldwide success of *The Da Vinci Code* and the latest book in the Harry Potter series is testament to the internationalization of the **publishing industry**. Both titles sold millions of copies, have been translated into numerous languages, and were made into feature films that became blockbusters and then sold millions of DVDs. Meanwhile a proliferation of small companies publish for niche markets. Consumers buy these books either through traditional means such as bookstores or over the Internet from outlets like Amazon.com, or even buy audio books and listen to them on their iPods.

Cultural and artistic environments also understand the concept and strength of **international and global branding**. Museums are actively participating in this trend. In March 2007 the Musée du Louvre in Paris announced an international collaboration with Abu Dhabi in the United Arab Emirates. The Louvre will lend its brand to the new museum for 30 years, for an estimated €400 million. It will also lend different exhibitions or art works to the new museum. Tate Britain has similar arrangements with Liverpool and St. Ives, while the Guggenheim Museum, based in New York City, has a presence in Venice, Bilbao, Berlin and Las Vegas.

These are only some of the trends and topics that we will cover in this chapter. The international arena for cultural industries is very broad and very active. The cultural industries are finding innovative ways and means of participating in this market. The next section examines the incentives and barriers that they face.

6.2 INCENTIVES AND BARRIERS TO INTERNATIONALIZATION

In most industries, from aerospace to digital technology, there are incentives to encourage internationalization. The cultural industries are no exception. The incentives invoked by the parties involved include saturation of their own market and access to additional or different resources.

Saturation of the local or national market is a strong incentive for a company to look elsewhere for growth. When the possibilities for growth are limited, the appeal of conquering foreign markets can be irresistible. The will or desire to access a growing market or develop a new one can be a vital part of a firm's strategy.

An export activity allows the firm to increase the profitability of an existing product that was conceived and developed for the local or national market. The company can take advantage of economies of scale or its extra production capacity.

Access to **additional financing** is another strong incentive. International collaboration or co-production has become an efficient tool for reducing financial risks for a given project and also for sharing the costs of a project. It is a means of accessing additional sources of financing in foreign markets. International collaboration can provide a company with access to foreign government subsidies that otherwise would be unavailable to it. It can also be an efficient way for smaller firms to develop partnerships with respect to large projects, in order to achieve economies of scale similar to their larger competitors.

In addition, international collaboration can be a means of **accessing technical, operational or marketing expertise,** allowing the firm to accelerate its own innovative cycle. Working Title Films, an independent film producer, was able to secure financing and access to international distribution by forming an alliance first with Polygram Film Entertainment and then with Universal Pictures.[6]

For major projects, the expertise of the partners involved could be a critical factor, for individually the firms might not be able to accomplish everything they have set out to do. This transfer of knowledge during the course of the project may benefit the firm in the short and medium terms. Acquiring such expertise might be possible only through international collaboration.

Another reason for internationalization might be **access to resources,** whether physical, material or human. For example, collaboration can provide access to geographical sites essential for shooting particular scenes in a movie. In some cases access would otherwise be too costly or simply impossible. Collaboration can also be a way of gaining access to certain actors, screenwriters or musicians.

On the other hand, different **barriers** might be in place to protect national cultural markets. These could include **regulations** or laws restricting access to distribution networks as well as controls on foreign ownership or copyright ownership.

Government **subsidy** can take various forms. It might take the form of fiscal measures, low interest rates, insurance coverage or non-refundable contributions. Often, a country's cultural industries are its most subsidized industries. Government subsidy can also come in the form of assistance with production, operations or foreign distribution. This form of subsidy can represent a very strong measure of protection. In Canada, for example, the cultural industries are supported through such government organizations as Telefilm Canada, the Canada Council for the Arts and Quebec's Société de développement des entreprises culturelles, thus giving cultural products a competitive edge. In the United States, aid to the cultural industries can come in the form of fiscal measures or donor contributions. Another example would be broadcast limitations on foreign programming, which reduce the air time available for foreign programs or movies.

Business practices are another form of protectionism. These can include use of local agents or local unions and even restrictions on foreign currency.

Finally, the **lack of protection of intellectual property** and the acceptance of pirating may be a barrier to internationalization. Pirating can take different forms, one of which is the unauthorized presentation or sale of a movie, music recording or television program in a hotel, bar, restaurant, museum or any other public place, including airing it on television. Pirating also includes the unauthorized reproduction of a book, DVD or CD.

6.3 FACTORS TO CONSIDER BEFORE ENGAGING IN INTERNATIONAL ACTIVITIES

Before deciding to export to or form a partnership in a foreign country, the company should analyze the possibilities, risks and challenges that could lie ahead. This evaluation will help the company to make wise choices, avoid costly mistakes and ultimately take advantage of any potential business opportunities.

6.3.1 Market Potential

The first step is to evaluate the potential of the foreign market. Libraries and the Internet are very good sources of information. National governments also publish pertinent statistics on the desired market. Most countries post online data relating to their cultural industries. For example, the European Audiovisual Observatory, which represents 36 states and the European Community, provides information on the audiovisual industry in Europe (http://www.obs.coe.int/index.html.en). A number of trade magazines publish information on the cultural industries and trends within each industry (http://www.variety.com; http://www.hollywoodreporter.com; http://www.billboard.com).

It is also useful to examine the general economic situation of the country of interest (purchasing power, strength of currency, etc.). The weekly magazine *The Economist* is one excellent source of information (http://www.economist.com). Demographic trends are another strong evaluator of market potential. Depending on the product or service one wishes to export, understanding rural or urban trends, population growth and life expectancy, and age density can be very useful.

6.3.2 Access

It is also important to determine whether the target country has any entry barriers such as restrictions or quotas on imports. Even restrictions on foreign currency could make it difficult to access that market.

Some consideration should be given to possible difficulties with piracy or the use of intellectual property without permission.

Finally, national subsidies will have to be considered. If the target country does subsidize its domestic market, it will be that much more difficult for foreign companies to compete in that market.

Access also refers to the geography of the target territory. The climate, the infrastructure of the country in terms of roads, and access to different population groups have to be considered. The existence of distribution channels is primordial – exporting books, for example, to regions where no distribution channels exist could become a very expensive proposition.

6.3.3 Competition

An evaluation of the intensity and quality of the competitive environment should be carried out. Discussions with embassy or consular representatives can be a good starting point.

The evaluation should determine what products are currently being offered by competitors or are likely to be offered in the future. Some consideration should be given to substitute products. If the target market is already saturated, then one should question the potential for success.

The strength of the competitors needs to be determined: their financial means and their market share.

6.3.4 Product Fit

The product or service that the firm wishes to export must be able to meet consumer expectations. If the product needs to be adapted to the different markets, an evaluation of the related costs should be done. Shipping and handling expenses and import restrictions will have to be considered.

It is very important for the company to become knowledgeable about consumer behaviour in the target market. In the United States, for example, the four major broadcasters (CBS, NBC, ABC and FOX) never air programs that are dubbed or subtitled.

Knowledge of the national culture is also crucial. For example, firms exporting to Arab countries should be aware of any restrictions that the national broadcaster or theatres might impose on the product; some images or scenes may need to be edited as they might be shocking to a different culture.

6.3.5 Service or Consultation after the Sale

If the product requires follow-up after the sale, this will have to be factored in. For example, when a television format is sold, production follow-up is required. There might be a need for travel to the different sites in order to provide consulting services. Quality control, or exercising control over the image or reputation of the product or service after the sale, can be important. Questions such as how the promotion will be carried out and where the product will be sold will have to be considered.

6.3.6 International Markets and Fairs

A good way to become acquainted with an international industry is to attend an international market or festival. Such events are held for most types of cultural product (MIDEM for music, the Cannes film festival for cinema, MIP-TV or Mipcom for television, the Frankfurt book fair for publishing).

6.3.7 A Visit to the Identified Territory

It is good strategy to visit the country of interest and meet with local industry representatives. This can be somewhat costly, but if the visit is planned properly it could result in valuable information that will increase the chances of success.

6.4 REVISITING THE MARKETING MIX

6.4.1 The Product

The product or service that the firm wishes to export must be able to meet consumer expectations. This leads to a major decision: whether to **adapt the product or standardize it**.

Adaptation might be required because of different languages, cultural values, tastes or even legal requirements. It is important for the company to fully understand consumer and market behaviour in the target market. For example, in the United States, since the four major broadcasters (CBS, NBC, ABC and FOX) never air movies or programs that are dubbed or subtitled, any attempt to sell a foreign-language film to one of these networks might be a waste of time and money. When foreign films are sold in the U.S. market very often they are adapted and produced using American actors. There *is* a market for foreign films in the United States, but it is limited to film festivals and repertory cinemas in some of the larger cities.

Understanding the national culture is crucial. When exporting to Asian countries, for example, the firm should be aware of the restrictions that theatres, publishers or the national broadcaster might impose on the product; some images, scenes or topics might need to be edited because they are considered shocking to a different culture.

An Oscar-winning film, *The Barbarian Invasions,* a France/Canada co-production, was re-edited for the international market because some of the content was considered too "local." Scenes about the health-care system, cable system and media concentration in Canada – 11 minutes in all – were cut. The film's co-producers, Cinémaginaire and Pyramide Productions, and its director, Denys Arcand, all agreed to adapt the film for the international market. However, both versions are available in Canada.

In adapting the product for the international market, the firm should concentrate on transferring quality – exporting the same perceived value as that offered in the domestic market. The firm should also consider the **actual use** of the product. A 2002 article in *The Economist* notes that one fifth of J. K. Rowling's Harry Potter books sold in Japan are in English: The books are being read not only by children but also by adults wishing to improve their English-language skills.[7]

The product or service exported might also be a licence: a licence to publish a book in a different language, to produce a television program from an existing concept, to produce a feature film from an existing film or book, or even to produce merchandising products. The U.S. television show *Survivor* is an adaptation of a European program, the film *Three Men and a Baby* is an adaptation of the French film *Trois hommes et un couffin,* and any number of feature films are based on novels – for instance, the Oscar-nominated film *Atonement* is based on Ian McEwan's novel of the same name.

The firm may decide to develop new products for specific foreign markets. This decision will involve additional costs – and risks, since one of the advantages of exporting a standardized product is the low cost.

Then there is the issue of **branding strategy**. The firm needs to make decisions regarding the brand. It can adapt the name and change it or create a new name. Each of these decisions entails different considerations and will have different consequences. The decisions might be influenced by the brand's reputation and by the need to adapt the product.

Harlequin Enterprises Limited is a Canadian book publisher founded in the early 1950s. It publishes some 110 fiction titles per month, in 27 languages in 95 international markets over six continents. The company has a pool of 1,300 authors. In 2004 it sold 130 million books.

Harlequin perceives itself as an expert in terms of branding strategy in the publishing industry. The company develops brands that correspond globally to the tastes of its readers. It offers readers an array of romantic storylines (traditional, suspense, inspirational, etc.) in a wide selection of brands linked to each theme. These include Harlequin Books, MIRA Books, Red Dress Ink, LUNA Books, HQN Books, Steeple Hill Books and Steeple Hill Café. In 2003, 12 Harlequin titles made it onto the *New York Times* bestseller list.

In the music industry, the name of the artist can be the brand. Céline Dion, who is globally renowned, has managed her brand and repositioned it as her career has evolved over the years. The repositioning was due first to her age, as she went from teenager to mature woman, then to the change in her repertoire, as she went from singing only in French to developing a career in English. Now she is known the world over mainly as Celine, and in most countries consumers actually think she is American rather than Canadian.

If the product requires follow-up after the sale, this will have to be planned. For example, when a television format is sold, production follow-up is usually required. There might be a need for travel to the various sites in order to provide consulting services.

6.4.2 Communications and Promotion Decisions

Just as in its domestic market, the firm will have to make decisions about which communications tools to use. As with the product itself, the message can be either adapted or standardized. One of the first considerations is **language.** The company must ensure that any translation is accurate. There might also be government regulations with respect to what is acceptable as a message.

Use of the same media as in the domestic market may be inappropriate or inadequate. A thorough knowledge of the **structure of media** in the target country, as well as the various levels of media penetration, is essential.

Competitor behaviour is always a good source of market information. In this regard, the Internet, **public relations** and word of mouth can all be used to good advantage. Any new communications trends should be evaluated. For example, product placement can be very effective in certain markets.

Another key decision is the choice of advertising agency: international or local. This choice will be based on many criteria, ranging from the quality and reputation of the agency, to the level of control the company wishes to maintain, to the capability of the agency to oversee international coordination of the company's message.

The company will have to address several issues: the message, whether adapted or not; the choice of media; and the budget for each territory, in light of the forecast sales or revenues for each territory.

6.4.3 Pricing Issues

Pricing issues will have to be addressed first in terms of costing. Any adaptation of the product or the communications tools will increase the original cost. Shipping expenses, import restrictions and the cost of preparing any informational materials should all be taken into account.

The factors to be considered will vary according to the objectives of the company. Is the main objective to penetrate the market? What is the expected profit? Foreign demand will also have to be addressed. What is the forecast demand? How quickly does the company wish to recuperate its investment or its costs?

The company will have to look at competitive or substitute products. The economic situation of the territory will also be a factor in price determination. The company should also look into the availability of subsidies from the national government. This will lower the financial risks and also may make the whole endeavour more interesting.

6.4.4 Distribution or Exportation Process

The final aspect to be addressed is distribution. How involved in distribution does the firm wish to be? What are the different means available to export a product or internationalize a firm? What should be exported – the product itself or simply a licence?

The distribution process varies from no involvement with limited risk and no control to total involvement with high risk and complete control. The process can also be part of the growth of the activity. The steps can be identified as follows:

- indirect export
- direct export
- use of an agent
- use of a distributor
- creation of an internal distribution department
- creation of a foreign distribution office

With **indirect export,** the firm is not involved in the process. The company's main activity remains local or national – its products are bought and resold to some other territory. Indirect export is possible only when copyright is not a factor – for example, with handcrafted jewellery made for the local market.

When the firm wishes to become involved, it may first consider subcontracting its entire distribution activity to an outside company. A **distributor** specializes in selling many similar products offered by a variety of firms. It is responsible for marketing activities and will normally work on a percentage-of-sales basis and deduct any promotional or marketing costs. If the product is particularly appealing, the distributor might agree to an advance payment, which may or may not be guaranteed. A distributor may either deal directly with many territories or use other distributors for specific territories. A distributor's strength resides in the quality and variety of the products that it represents.

Since the distributor represents many products and therefore many clients, its representation of the product will be somewhat diluted. Also, by using a distributor the producer foregoes the possibility of learning about international markets and developing direct contact with international clients. However, using a distributor is considerably less costly to the producer than selling directly.

The firm may also wish to use one or more **agents**. An agent is an independent representative working either in a single territory or in several territories. The agent is not exclusive to the firm, represents several clients and has a minimal structure.

Usually, the agent gets paid a percentage of the gross sale and is not responsible for any marketing expenses incurred. The agent acts as an intermediary or consultant, does not guarantee any revenue and does not give any advances.

The firm may then decide to create an **internal distribution department**. This can be quite costly. The company will need to hire salespeople, do its own market analyses and build its own international contacts. To justify the cost of these activities, production and sales volume will have to be high.

Direct distribution permits the company to exercise much greater control over the products being distributed. It eliminates distribution costs and facilitates the development of international contacts. If the firm's strategy is to internationalize, direct distribution will enable it to become acquainted with the international market and to cultivate solid contacts leading to strong relationships.

The distribution department may eventually locate in the foreign territory. A **foreign distribution office** will allow the firm to maintain a presence in crucial markets and to be much more competitive within the industry.

The firm will probably choose a **mixture** of all of these different distribution arrangements. Depending on the territories to be served, it might choose a distributor in South America, hire an agent in the United States and distribute directly in a territory where it wishes to build particularly strong relationships.

6.4.5 Digital Distribution

The traditional systems for distributing DVDs, CDs and books are at a critical juncture, especially with the proliferation of Internet services, which has become both an opportunity and a challenge. Consumers are buying their music, books and DVDs online. This has led to the availability of products that otherwise would not be accessible to consumers because of the high cost of shelf space to retailers. The average outlet in the Barnes & Nobles bookstore chain carries 130,000 titles. Yet more than half of Amazon.com's book sales come from titles outside its top 130,000 titles.[8]

The volume of online sales is increasing year by year. Chris Anderson, in "The Long Tail," describes this phenomenon in detail, and the advantages for both the producer and the consumer. This new economic model for the media and entertainment industries provides for an unlimited selection of goods. Consumers are scouring online catalogues and are discovering titles that might no longer be available in stores. Even megastars are opting for this mode of distribution. For example, Paul McCartney has abandoned EMI Records in favour of iTunes so that his music can be sold on the Internet.

6.5 JOINT VENTURE AND DIRECT INVESTMENT

6.5.1 Collaboration: From Co-production to Joint Venture

More than ever before, cultural industries are using co-production as a means of becoming internationalized. This type of collaboration allows the partners from different countries to be involved in a given project not only financially but creatively as well.

In television and film production, many countries are negotiating bilateral agreements whereby a co-production qualifies as national content for each country even if it is not a majority partner.

Why co-produce? There are many reasons why producers choose co-production:

- Access to new markets. International co-productions provide access to markets that otherwise might not be available.
- Government protection. Because of restrictions imposed by certain governments, access to the market or resources is possible only through international collaboration.
- Sharing of financial risks. Co-production allows for the sharing of risks on major projects.
- Economy of scale. On large projects, collaboration permits economy of scale in different aspects of production, thus allowing the partners to compete within the industry.
- Transfer of know-how. Co-production facilitates the exchange of knowledge and skills between the partners.

Co-production: minority or majority? This decision will depend on the firm's resources and objectives.

Feature films have been using this mode of production for years. Many blockbusters, such as *The Lord of the Rings,* are co-productions. Likewise, many documentaries are co-productions between the United Kingdom and Canada or the United Kingdom and the United States. In the publishing industry, co-publishing allows for lower unit production costs. Museums are now co-producing exhibitions that tour from one museum to the other. The musical *Notre-Dame de Paris* is a co-production between France and Canada.

However, co-production has its **disadvantages**.

Coordination costs are usually higher since the project will be more complex. There is some loss of control over content and also a loss of cultural specificity. The foreign partner may wish to take advantage of the situation and thus can become a competitor instead of a partner. Furthermore, partnerships are not easy to manage. The objectives of one partner might not match those of the other partner.

Experience has shown that certain factors contribute to the success of a partnership.

The choice of partner is the main consideration. It is essential that there be coherence between the choice of partner and the objectives of the project. A company should never form a partnership with the first candidate that appears on the scene.

The relationship between the partners will greatly influence the atmosphere of the project and ultimately its success. The partners should have complementary assets and synergies. If they are different in terms of size, at least their objectives should be compatible. They should have a similar corporate culture, and their relationship should be flexible, committed and based on trust.

The project leader, or manager, should be experienced, open to different cultures and capable of working in a turbulent environment, in a matrix-type structure. He or she must be a communicator and a diplomat, able to cultivate a climate of trust and to meet the objectives of each partner.

The members of the team must be able to work autonomously in a semi-structured, complex environment. Their tasks will be non-repetitive. They will have to work in teams, keep pace with their colleagues, and focus on the process and the structure. They must possess intuition, a tolerance for ambiguity and a high degree of self-confidence.

From the start, there should be a clear, precise, **detailed contract,** in order to avoid any future disagreement or confusion. The contract should specify the different roles and responsibilities and should also contain clauses regarding conflict resolution.

6.5.2 Foreign Direct Investment

This is the activity with the highest risk yet the one that yields the most control. Its objective is to develop long-term interest in a territory or in a foreign company.

The incentives for a firm to invest might have to do with marketing factors, either restrictions or opportunities. The restrictions might be imposed by the government, and the firm might need to invest in order to increase its activities. The opportunities are many, from access to emerging markets to the export and control of particular know-how.

Foreign direct investment might also be a way of reducing costs by using less expensive skilled labour. In addition, the investment climate in a particular territory or firm might be just right in terms of growth.

However, any investments should be carefully analyzed in advance. If the opportunities are many, the risks might be as well.

6.6 CONCLUSION

Throughout these activities, controls and performance measurement should be a focus of attention. Objectives should be clear and quantifiable, and results should be examined against them on a regular basis.

Although international markets are complex, in this day and age few companies can avoid them. The company should first become fully acquainted with the different business opportunities and then make a series of decisions in terms of adaptation or standardization.

These decisions must not be taken lightly, since a bad choice could have serious repercussions for the company not only internationally but domestically as well.

SUMMARY

International marketing is a complex activity that calls for planning and control. The company will have to develop an awareness of international trends and the various risks and opportunities involved. Market research is crucial. The company will have to reach decisions about whether to adapt or standardize. It will most likely end up with a mix so as to address each market in the most appropriate manner. As the activity progresses, the company might decide to go beyond simply exporting and will consider joint ventures or foreign direct investment. Each option has its strengths and drawbacks and must be considered carefully.

Notes

1. *Focus 2007: World Film Market Trends* (Cannes: Marché du Film, 2007); *Profile 2007: An Economic Report on the Canadian Film and Television Production Industry* (Ottawa: Canadian Film and Television Producers Association, 2007).
2. India statistics: *The Indian Entertainment and Media Industry Report 2006* (New Delhi: Federation of Indian Chambers of Commerce and Industry, 2006).
3. *Focus 2007: World Film Market Trends*, op. cit.
4. John Colapinto, "Onward and Upward with the Arts. When I'm Sixty-four: Paul McCartney Then and Now" (*New Yorker,* June 4, 2007), p. 56–67.
5. http://www.simpleplan.com/.
6. J. Brunet and G. Gornostaeva, "Working Title Films, Independent Producer: Internationalization of the Film Industry" (*International Journal of Arts Management*, Vol. 9, No. 1, 2006), p. 60–69.
7. "Potter's Japanese Adventure" (*The Economist*, February 14, 2002).
8. C. Anderson, "The Long Tail" (*Wired*, Vol. 12, No 10, 2004).
9. Ibid.

Marketing Plan Template

Mission, Vision and Values

Narrative describing the meaning of the organization's mission, vision and values to the marketing department.

Situation Analysis

Internal Environment

- Aesthetic/Creative
- Organizational/Operational
- Financial/Human Resource Capacity
- Audience/Programming Capacity
- Audience Participation: Past, Present and Future Trends
- Key Institutional Issues

External Environments

- Cultural
- Sociodemographic
- Economic
- Political
- Educational
- Technological
- Geographic

Allies and Competitors

The Market

Target markets		Patron characteristics							Program characteristics						strengths	weaknesses	opportunities	threats
		age	family status/ size	gender	educational attainment	geography	lifestyle	cultural preferences	subscriber or member	single ticket or individual admission	lapsed subscriber or member	potential single ticket/ individual admission	cost-sensitive patron	status-seeker				
Existing	mkt A																	
	mkt B																	
New	mkt C																	
	mkt D																	

Strategies

Advertising strategies
    rationale: value and potential impact of each strategy
Branding/positioning strategies
    rationale: value and potential impact of each strategy
Packaging strategies
    rationale: value and potential impact of each strategy
Pricing strategies
    rationale: value and potential impact of each strategy
Promotional strategies
    rationale: value and potential impact of each strategy
Publicity strategies
    rationale: value and potential impact of each strategy
Target market strategies
    rationale: value and potential impact of each strategy

Objectives

	target market	sales goal	revenue goal	participation objective(s)	current position within purchasing process
subscriptions/memberships					
package A	mkt A				
	mkt B				
	mkt C				
	mkt D				
package B	mkt A				
	mkt B				
	mkt C				
	mkt D				
performances/exhibitions					
perf/exh A	mkt A				
	mkt B				
	mkt C				
	mkt D				
perf/exh B	mkt A				
	mkt B				
	mkt C				
	mkt D				
related programs/events					
program A	mkt A				
	mkt B				
	mkt C				
	mkt D				
program B	mkt A				
	mkt B				
	mkt C				
	mkt D				
special services/amenities					
service A	mkt A				
	mkt B				
	mkt C				
	mkt D				
service B	mkt A				
	mkt B				
	mkt C				
	mkt D				

Tactics

Tactic	medium (media)	target market(s)	key message(s)	strategy(ies) supported	timing			costs		goals		qualitative metrics
					development	implementation	completion or delivery	financial	human resources	sales	revenue	
subscriptions/memberships												
Tactic A												
Tactic B												
performances/exhibitions												
Tactic C												
Tactic D												
related programs/events												
Tactic E												
Tactic F												
special services/amenities												
Tactic G												
Tactic H												

Index

N

O

P

R

S

T

U

V